GUIDE TO THE

MANTA & DEVIL RAYS

OF THE WORLD

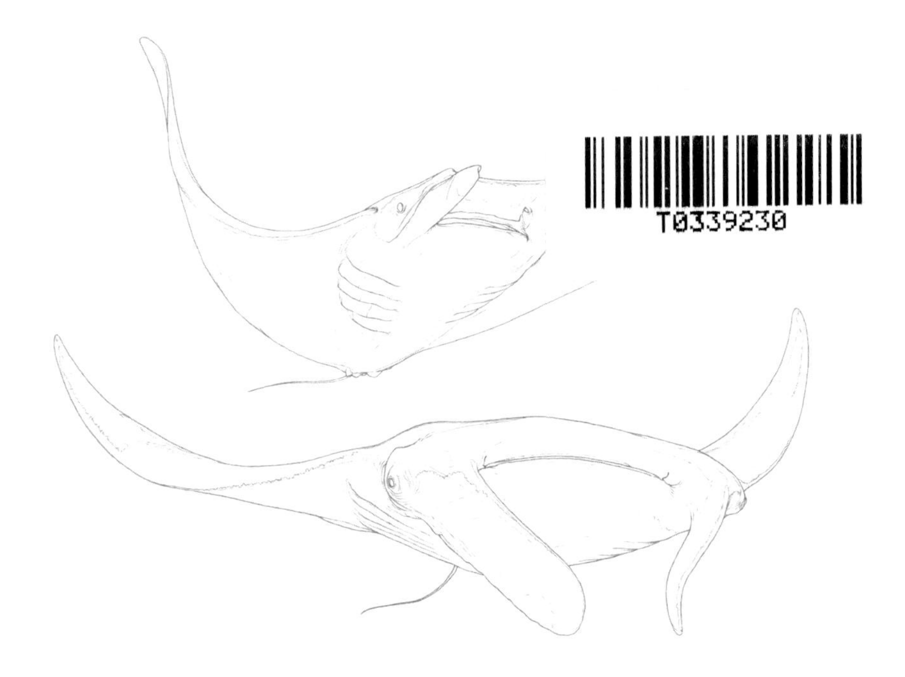

Guy Stevens, Daniel Fernando, Marc Dando
and Giuseppe Notarbartolo di Sciara

Royalties from the sale of every copy go to The Manta Trust

Production of this book was partly sponsored by Save Our Seas Foundation

Published by Princeton University Press,
41 William Street,
Princeton, New Jersey, 08540
press.princeton.edu

Published in the United Kingdom and the European Union by
Wild Nature Press

Library of Congress Control Number: 2018935422
ISBN: 978-0-691-18332-9

British Library Cataloging-in-Publication Data is available

Printed in Slovakia

10 9 8 7 6 5 4 3 2

Front cover: above – Sicklefin Devil Ray *Mobula tarapacana*, below – Oceanic Manta Ray *Mobula birostris*

Back cover: a school of Brassy Chub *Kyphosus vaigiensis* scatters as a hungry Reef Manta Ray *Mobula alfredi* powers its way through the surface waters of Hanifaru Bay in the Maldives

CONTENTS

Acknowledgements

The authors of this guide would like to thank the Manta Trust team across the world and its scientific collaborators and advisors who provided vital information, feedback, and constant encouragement to make this guide a possibility.

We would like to especially thank the Save Our Seas Foundation (SOSF) for their generous support in funding this project in its entirety and for their continued partnership over the years. SOSF have placed their trust and confidence in the Manta Trust since its inception and for that we are very grateful. We also extend a special thanks to current CEO Michael Scholl for believing in this project and to the rest of the SOSF team for their support, in particular Sarah Fowler and Nadia Bruyndonckx.

We also appreciate all the incredibly talented photographers who have very graciously donated their images for this important project (see photo credits below).

Finally, we express our thanks to all those who contributed towards this guide, their valuable insights, and unfettered enthusiasm. Our sincerest apologies to anyone we might have erroneously left out.

Photographic credits

Shannon Arnold 121B
Frédéric Bassemayousse / WWF 91B
Lisa Bauer 121C
Calvin Beale 79B
Roxane Borruat 123B
Adam Broadbent / Scubazoo Images 119B
Tom Burd / tomburd.co.uk 10, 16T, 131T
Richard Carey 29C
Bex Carter 133, 134
Danny Copeland 103B, 121T
Mike Couffer 90T
Ryan Daly 33B
Framoudou Doumbouya 98T
Alexandra Fennell 34T
Daniel Fernando 115C (both), 116T, 118C, 140TL
Chiara Fumagalli 41B
Shawn Heinrichs / Blue Sphere Media LLC 99B, 110T, 115T, 116B
Erick Higuera / erickhiguera.com 127B
Simon Hilbourne 87B, 138B
Paul Hilton / paulhiltonphotography.com 117, 118R, 139BR, 140 all except TL, 141 all
Emily Humble 17B
Jason Isley / Scubazoo Images 107B
Gisela Kaufmann 139TR
Ken Lancaster 20C
Jeff Lemelin 23B
Andrea Marshall 86T
Moosa Mohamed 36B
Rory Moore 66T
Andy Murch / elasmodiver.com 82T
Duncan Murrell / duncanmurrell.com 47BR
Alex Mustard / amustard.com 83B
Michael Nolan / mikenphotography.com 15T, 95B
Parnupong Norasethkamol 78T
Doug Perrine 46T
Thomas P. Peschak 118T, 118BL, 139BL
Andy Sallmon / seait.com 111B
Tam Sawers 36C
Alessia Scuderi / Tethys 26B
Marc Sentis 106T
Tane Sinclair-Taylor / tanesinclair-taylor.com 39B, 102T
Guy Stevens 13B, 14T, 15B, 19T, 19B, 20T, 20B, 21T, 21B, 22T, 22B, 23T, 30T, 31B, 32B, 33T, 34B, 35T, 35B, 36T, 37T, 37B, 38B, 39T, 40, 43T, 43B, 46B, 47T, 47C, 47BL, 48T, 49, 50, 52T, 52B, 53T, 62T, 64T, 64B, 70T, 71B, 74T, 75B, 94T, 112, 114T, 114BL, 114BR, 115B, 120T, 122T, 122C, 126T, 128T, 128B, 129T, 129B, 130T, 131B, 132T, 132B, 139TL, 142, back cover
Vittoriano Ummarino 28B
David Valencia / david-valencia.com 32T
Joost van Uffelen / joostvanuffelen.com 42B
Sergey Uryadnikov / Shutterstock 29B (both)
Mohamed Visham 114C

The authors

Guy Stevens

Guy Stevens is the Chief Executive and Co-Founder of the Manta Trust, a UK and US registered charity dedicated to the conservation of manta rays, with collaborative projects in over 25 countries. He has spent the last 15 years studying manta and devil rays all over the world and is one of the foremost experts on these species. After completing a degree in Marine Biology and Coastal Ecology at the University of Plymouth in the UK, Guy moved to the Maldives, where he founded the Maldivian Manta Ray Project in 2005. After 11 years of research, Guy completed his PhD on the world's largest population of reef manta rays at the end of 2016. Guy's conservation efforts in the Maldives have led to the creation of several marine protected areas at key manta aggregations sites, most notably at Hanifaru Bay. Internationally, Guy is part of a team which has driven the conservation of mobulids forward, resulting in the listing of all manta and devil rays in the Appendices of the Convention on the Conservation of Migratory Species of Wild Animals (the Bonn Convention) and the Convention on International Trade in Endangered Species of Wild Fauna and Flora (CITES). In recognition of his contribution to ocean conservation, in 2015 Guy received an Ocean Award for his work on manta rays, and in 2017 he was awarded the UK Prime Minister's Points of Light Award. Guy is the author of the world's first book on manta rays; *MANTA – The Secret Life of Devil Rays*, published in 2017.

For more information, visit **www.mantatrust.org**

Daniel Fernando

Daniel Fernando is an Associate Director of the Manta Trust, and Co-Founder of Blue Resources Trust – a Sri Lankan NGO dedicated to promoting informed, science-based decision making to facilitate the conservation and sustainable use of marine resources, and to provide a platform for scientists to expand research in the Indian Ocean. Daniel has spent over six years investigating mobulid ray fisheries as part of his PhD research, and has helped set up data collection projects all across the world that provide essential information to increase understanding of the extent of overexploitation of these vulnerable species. In addition to field research, his drive to encourage a shift toward sustainable fisheries has resulted in his appointment as an Advisor to the Minister of Sustainable Development and Wildlife of Sri Lanka, where he represents the Government of Sri Lanka and drives policy measures forward through international conservation instruments such as the Convention on the International Trade in Endangered Species and the Convention on the Conservation of Migratory Species of Wild Animals. He is an Marine Conservation Action Fund (MCAF) Fellow and member of the IUCN Shark Specialist Group.

For more information, visit **www.blueresources.org**

Marc Dando

Marc Dando is a scientific illustrator and publisher. He has always balanced art and science, completing an honours degree in Zoology at Nottingham University then moving into a career as a graphic designer. After many successful years in commercial graphics, an opportunity arose to work on various natural history projects, initially in design, but finally in scientific illustration.

For the past 25 years he has worked alongside many eminent scientists, especially in the field of marine biology. His first book, *Sealife: a complete guide to the marine environment,* was as a co-author, designer and illustrator; but it was *Sharks of the World,* a three-and-a-half-year project with Leonard Compagno and Sarah Fowler, that established him as an outstanding scientific illustrator. With Dave Ebert he has now worked on the updated *Sharks of the World* and has collaborated with him on further projects for the FAO and IFAW. Although his work has been seen in Monaco and The Mall Galleries in London, the majority of his work is found in books, field guide literature and magazines.

Giuseppe Notarbartolo di Sciara

Giuseppe Notarbartolo di Sciara is an Italian marine conservation ecologist who has worked for 40 years to advance knowledge of the natural history, ecology, behaviour and taxonomy of marine mammals and cartilaginous fishes. He earned a PhD degree in marine biology at the Scripps Institution of Oceanography (University of California at San Diego) in 1985 with a thesis on the taxonomy and ecology of devil rays, describing a new species, *Mobula munkiana*. In 1986 he funded the Milan-based Tethys Research Institute, which he chaired and directed until 1997 and again from 2010 to 2016. In 1991 he spearheaded the creation of the first high-seas marine protected area, the Pelagos Sanctuary for Mediterranean Marine Mammals, established in 1999 by a treaty amongst Italy, France and Monaco. He has served as the Italian Commissioner at the International Whaling Commission (1999–2004), and as Chair of the Scientific Committee of ACCOBAMS (2002–2010). Currently the CoP-appointed Councillor for aquatic mammals at the Convention on Migratory Species; co-chair of the IUCN Task Force on marine mammal protected areas; deputy chair of the IUCN Cetacean Specialist Group; member of the IUCN Shark Specialist Group; regional coordinator for the Mediterranean and Black Seas of IUCN WCPA – Marine; and Advisor, Pew Fellows in Marine Conservation. He has lectured in science and policy of the conservation of marine biodiversity at the University Statale of Milan from 2007 to 2016. Author of over 160 scientific publications, several books and popular works.

For more information, visit **www.disciara.org**

Save Our Seas Foundation

In 2007, the Manta Trust's research programme in the Maldives became one of the first conservation projects to be supported by the Save Our Seas Foundation (SOSF), and in the intervening decade this collaboration has evolved into a proud partnership. With funding and guidance from the SOSF and the authors' passionate and enduring dedication to manta and devil rays, the Manta Trust has grown into a highly influential organisation at the forefront of global mobulid conservation. We are delighted to support the creation of this field guide, which highlights the great advances in science and conservation the Trust has achieved since its inception.

The Save Our Seas Foundation has funded more than 200 projects in over 50 countries around the globe. Each project strives for deeper understanding and more innovative solutions in marine research, conservation and education. With more than a decade of experience notched up, we look forward to the next decade as an ongoing process of evolving and refining. We continue to grow and learn as a foundation, seeking out and supporting the best and brightest people whose innovative projects make a real and lasting impact for the health of our oceans – and ultimately for every person on the planet. We share science beyond the boundaries of scientific publications through online, print and multimedia channels.

RESEARCH Good science is essential because in order to conserve anything, you must begin by understanding it.

EDUCATION Learning yields understanding and inspiration, which give rise to action and ultimately lead to change.

CONSERVATION We support projects that will have a direct, measurable impact for conservation at a local or global scale.

SPECIALISATION We have carved out a niche in the world of shark and ray conservation.

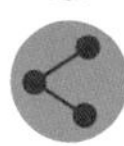

COMMUNICATION We share science beyond the boundaries of scientific publications through online, print and multimedia channels.

CULTIVATION Nurturing young people and early career professionals, whom we prioritise through various funding opportunities, will sustain ocean conservation in the long term.

INNOVATION We look for innovative approaches and new technology that will help achieve research, conservation and education gains.

COMMITMENT We offer long-term support to a few select conservation projects through partnerships and to the Save Our Seas Foundation research and education centres around the world.

COLLABORATION We work closely with our project leaders during their funding period to encourage them and to facilitate collaboration between them.

The Manta Trust

The Manta Trust is a UK- and US-registered charity (US = Action for Mantas) that co-ordinates global research and conservation efforts for mobulids, to better protect these vulnerable animals and the habitats on which they depend. Since its formation in 2011 with projects in five locations, the Manta Trust has grown in reputation and size, developing robust relationships at local, regional, and national levels in over 25 mobulid range states. Our international network is comprised of a diverse group of researchers, scientists, conservationists, educators and media experts.

Our vision is a sustainable future for the oceans, where mobulids and their relatives thrive in healthy, diverse marine ecosystems. The Manta Trust's unique, multifaceted approach to achieving this vision enables us to maximise the conservation opportunities. We bring together international expertise on mobulid rays and marine ecosystems to achieve effective, global conservation solutions. We also utilise the manta ray's charisma as a flagship species to promote and engage people with wider marine conservation issues. In this way, mobulids have become a catalyst for change; engaging and motivating the public, governments and local communities alike with the preservation of ocean life more generally.

We work with partner NGOs and scientific institutions to examine the spatial ecology, population dynamics and susceptibility of manta and devil ray populations to mankind's impact on the planet, while developing real-world strategies for public involvement in mobulid ray protection, ecotourism and management. We also consider people in our approach. Citizen science and the engagement of the general public in our work is a large part of what we do, from encouraging the submission of photo identification images and making our results publicly available, to writing about our work in books and magazines, and using our social media outlets to engage our audience in our research. We encourage volunteers and students to participate in our programmes. Most importantly, we always partner with local initiatives to ensure that any practices we put into place have lasting results for the wider communities in which we work.

Since its inception, the Manta Trust's global network has positioned itself as the leading authority on mobulids. We are proud of what we and our colleagues have achieved in just a few short years as a global player in marine conservation. Some of our key collaborative achievements to date include: gaining Appendix II listing for all mobulids on the Convention on International Trade in Endangered Species; gaining Appendix I and II listing for all mobulids on the Convention on the Conservation of Migratory Species of Wild Animals; developing, in collaboration with WWF and Project AWARE, the first scientifically advised Best Practice Guidelines for Shark and Ray Tourism; assisting the IUCN Shark Specialist Group in developing a Global Conservation Strategy for Manta and Devil Rays; and publishing *MANTA: Secret Life of Devil Rays*, the world's first book dedicated to manta biology. The Manta Team has also played a key role in gaining national protection for manta rays in the Republic of Maldives, Indonesia and Peru.

INTRODUCTION

Left: a Sicklefin Devil Ray *Mobula tarapacana* cruises around a seamount in the Atlantic Ocean off the coast of the Azores.

An introduction to manta and devil rays

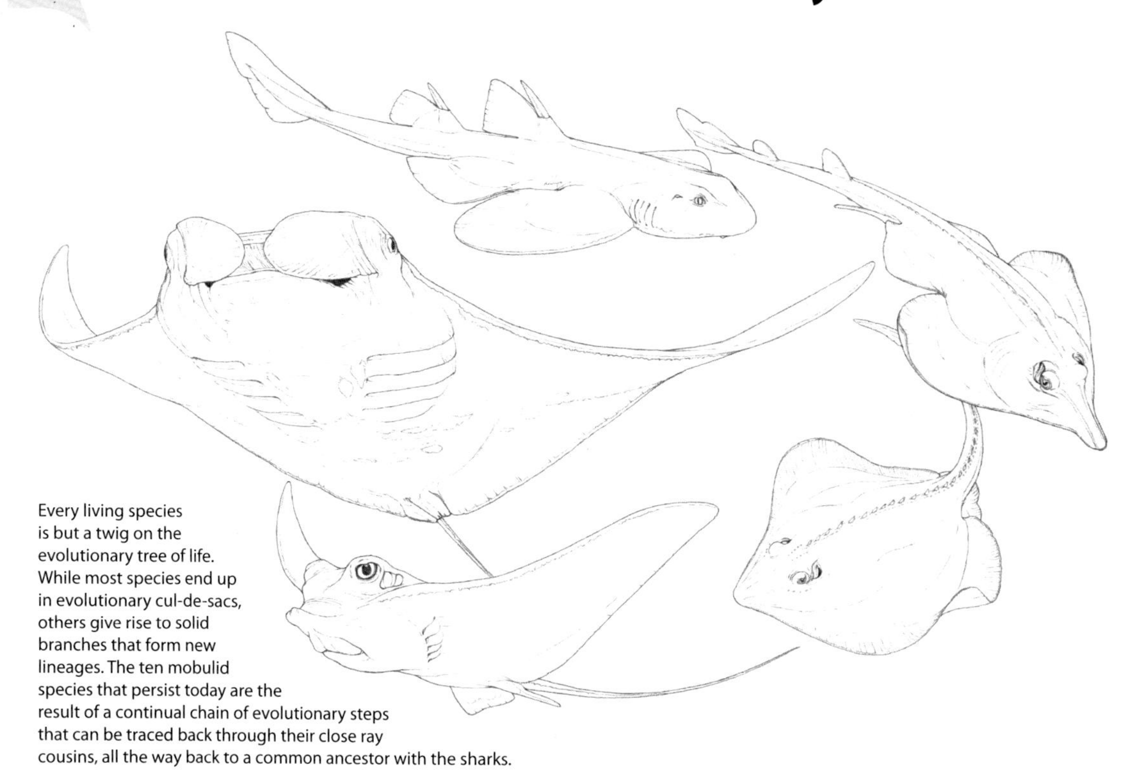

Every living species is but a twig on the evolutionary tree of life. While most species end up in evolutionary cul-de-sacs, others give rise to solid branches that form new lineages. The ten mobulid species that persist today are the result of a continual chain of evolutionary steps that can be traced back through their close ray cousins, all the way back to a common ancestor with the sharks.

Manta and devil rays, collectively known as mobulids, like all sharks and rays are classified as elasmobranchs – cartilaginous fishes that belong to one of the two subclasses within the Chondrichthyan fish class (the other subclass is comprised of around 45 species of wonderfully strange looking fishes known as chimaeras). Currently the 1,160 or so species of extant (living) elasmobranchs are divided roughly in half, with about 510 species of sharks and 650 species of rays.

Somewhat resembling modern skates, the first rays appeared in the oceans approximately 150 million years ago, radiating from a common ancestor with the sharks. Their flattened body shape is essentially a squashed version of the archetypal shark, with internal physiology very similar to that of their cousins. There are nine extant orders of sharks, and four extant ray orders: Rajiformes (skates), Rhinopristiformes (guitarfishes and sawfishes), Torpediniformes (electric rays) and Myliobatiformes (stingrays and relatives, including mantas and devil rays).

Mobulids are taxonomically linked to all elasmobranchs by their similar skeletons, comprised of flexible, fibrous and light cartilage, as opposed to the dense bony skeletons of the vast majority of all other fish (approximately 30,000 species) and terrestrial vertebrates. This lightweight skeleton is about half the density of bone, saving valuable energy for the mobulids as they swim through the water. Unlike most free-swimming bony fishes, mobulids and their relatives have not needed to evolve a gas-filled swim bladder to compensate for a denser bony skeleton. Enlarged and extra oily livers compensate for their weight, but overall they are still negatively buoyant and will begin to sink slowly if they stop swimming. For some species, including the manta and devil rays, this leads to a life of perpetual forward motion creating lift.

While some sharks and rays spend their whole life in motion, the vast majority of ray species, and a large proportion of shark species, spend a great deal of time resting on the seabed. They can do this without asphyxiating by actively pumping water over their gills, much as the majority of bony fishes do. Because nearly all sharks and rays have their mouths on the underside of their heads and not in front, they have evolved small openings called spiracles, just behind the eyes, on either side of their heads. These spiracles allow the animals to draw in clean water without sucking up the sediment beneath them through their mouths. Mobulids still retain these spiracles, although there no longer appears to be any functional use for them in these constantly swimming species. In fact, the spiracles often house hitchhikers – small remoras that use these safe hidey-holes to live and take shelter in. On the manta rays, small flaps of skin cover the spiracles, making it hard to discern their presence just back from their eyes on the top of the ray's head.

Like all species that live life continually on the move in open water, mobulids have no need to actively pump water over their gills because the animals' constant forward motion pushes all the oxygenated water they need in through their mouths and over their gills. These 'obligate ram ventilators' soon asphyxiate and die if prevented from swimming. This would rarely occur under natural conditions, but is unfortunately often the case when they become entangled in fishing nets or fishing lines.

Mobulids are relatively large, slow-growing, migratory animals that form small, highly dispersed populations. They are among the least fecund of all sharks and rays, giving birth to a single pup every two to seven years after a gestation period of about one year. Such life history characteristics make them among the least productive species in the ocean and therefore very susceptible to anthropogenic pressures. Their aggregatory nature and schooling behaviour can make them vulnerable to even artisanal fisheries.

Growing global demand for the dried gill plates of mobulids in Asian medicine has exacerbated this issue and led to an increase in target fishing, along with higher retention in bycatch fisheries. This unsustainable demand has resulted in severe population declines across their range which, combined with their limited ability to recover from a state of depletion, greatly threatens their survival.

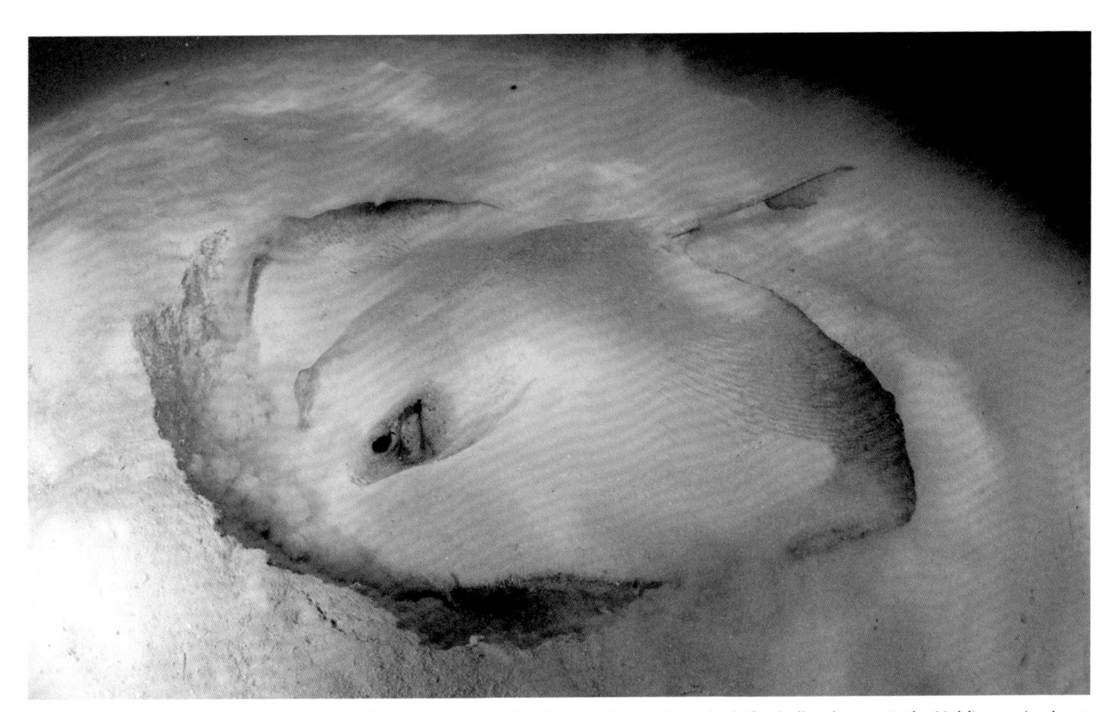

A Feathertail Stingray *Pastinachus sephen* digs for crustaceans and molluscs in the sandy seabed of a shallow lagoon in the Maldives, using large spiracle openings behind each eye to breathe in clear water while its mouth is buried in the sand. Although the stingray's manta and devil ray cousins still retain these spiracles, a life of constant motion in the open water column means these breathing holes are greatly reduced in size and function.

Taxonomy and evolution

A giant mouth opens wide as a Reef Manta Ray *Mobula alfredi* hovers above a cleaning station in the Maldives. Like all mobulids, manta rays are filter feeders, using five pairs of modified gill plates which encircle their mouth and paddle-like cephalic fins (head fins) to strain zooplankton and small fishes from the water.

Manta and devil rays (collectively known as mobulids) belong to a group of rays called the Myliobatiformes, which contains 12 families and about 370 species. Mobulids (Mobulidae) are most closely related to the eagle rays (Myliobatidae and Aetobatidae) and cownose rays (Rhinopteridae). All are characterised by diamond-shaped bodies and wing-like pectoral fins which they use to propel themselves through open water. Eagle rays and cownose rays all feed on the seabed, using their mouths to dig among the substrate in search of buried molluscs and crustaceans. The manta and devil rays, however, have taken to a completely pelagic way of life, never resting on the seabed.

Taxonomically within the Mobulidae family there is just one genus: *Mobula*, which contains ten species – two (possibly three) manta species and eight devil ray species. All are filter feeders, using their mouths and modified gill plates to strain plankton and small fishes from the water. In general, devil rays are much smaller than the manta rays and can be distinguished by morphological differences in their mouths and cephalic fins (head fins). Devil rays have a bottom jaw that is undercut so that the edge of the lower jaw rests much further back than the upper when their mouths are closed, whereas manta rays' jaws are aligned evenly.

The other differentiating anatomical feature is the shape of the cephalic fins, which when rolled up look like horns projecting off their heads – hence the name 'devil rays'. The primary function of these fins is to help funnel planktonic food into the gaping mouths of the rays when they are feeding. Unfurled, the devil ray's cephalic fins are a little smaller than the manta's, which unravel to form large, paddle-like structures that touch in the centre, creating a complete funnel around the manta's mouth.

Very little is known about the devil rays. Unlike the mantas, they are generally very shy towards divers, making it hard for scientists to observe their behaviour in the wild. Like the mantas, they are found throughout tropical, subtropical and warm temperate oceans and seas; some species seasonally aggregate in vast shoals of many thousands, probably coming together to mate. These schooling species also seek safety in numbers.

The first true mobulids appeared in the fossil record during the Middle Oligocene epoch 28 million years ago. Manta and devil rays are considered to be among the most recently evolved and highly derived groups of all elasmobranchs. Their flattened body shape is superbly adapted for life in motion; they are extremely efficient swimmers, gliding effortlessly through the oceans in search of food and mates. With one powerful beat of their wing-like pectoral fins they are capable of short bursts of speed easily in excess of 33km/h (20mph). Their flattened shape also helps to protect against predators, which find it hard to bite hold of the mobulid rays in an area of the body that contains vital organs.

Powerful wing-like pectoral fins propel manta and devil rays through the water. Occasionally, when these fins break the water's surface, they are mistaken for the dorsal fins of sharks.

A Spotted Eagle Ray *Aetobatus ocellatus* shows off her mollusc and crustacean-crushing mouth to the camera. Close relatives of the manta and devil rays (mobulids), eagle rays also use powerful wing-like pectoral fins to swim through the water column. However, unlike the mobulids, they also rest on the seabed.

Mobulid genetics

Recent genetic studies suggest Oceanic Manta Rays *Mobula birostris* and Sicklefin Devil Rays *Mobula tarapacana* are very closely related within the mobulid ray family. At over a dozen locations around the world these two species have been documented socialising together. The interactions always involve the smaller devil rays trailing in a line behind the larger manta. The reason for these interactions remains unclear. However, given the social nature of these relatively intelligent species, it is likely that curiosity towards an unfamiliar yet similar animal is at play.

Genetic analysis is a powerful tool for species identification and can provide insights into both species origins and boundaries. As species diverge from a common ancestor, their genomes accumulate different genetic mutations through time. By assessing the similarity of DNA sequences across individuals, it is therefore possible to ask questions such as 'how recently have they diverged?' and 'do they constitute separate species?'. DNA sequence analysis also facilitates the assignment of species to samples of unknown origin.

At their most basic, genetic studies can take the form of DNA barcoding, which involves comparing a single gene across multiple individuals. At the more complex end of the spectrum, whole genome sequencing can enable comparison of the entire genetic code of multiple individuals. In between these two extremes, there are various options for sequencing random portions of a genome. Critically, these restricted approaches make it possible to sequence the same random portions across multiple individuals to allow for direct comparisons of DNA content.

As has been extensively documented, identifying mobulids to species level presents several challenges. First, mobulids show strong morphological similarities and exhibit substantial range overlap, which can result in confusion when identification is carried out in the field. Second, mobulids are often observed as incomplete specimens in fish markets, with diagnostic parts removed, making species identification even more challenging. Unsurprisingly this has led to the group being taxonomically ambiguous; *Mobula alfredi* was only recognised as a distinct species in 2009, and in 2017, a proposal was published to collapse *Manta* and *Mobula* into a single genus and for three species pairs to be recognised as single species. By utilising genetic tools, taxonomic ambiguities can be ironed out and individuals can be assigned to species based on the genetic code contained within each tiny cell.

The starting point for any genetic study is a biological sample from the organism of interest. This can take the form of blood, tissue, muscle, skin or any other part of an organism that contains its cells.

It is now incredibly quick, easy and cost effective to extract high quality DNA from a well-preserved sample, which can then be manipulated in the laboratory for sequencing. For mobulids, the most common sample type is skin and muscle tissue, no larger than a pea, either collected from dead individuals in fisheries, or as biopsies from live individuals.

However, when designing a genetic study, important considerations regarding tissue collection should be recognised. It relies on access to individuals to sample, and to materials with which to preserve the sample. In addition, due to the listing of mobulid rays on the Convention on the International Trade in Endangered Species (CITES), tissue samples require permits to be moved between countries for analysis, which can be time consuming. Furthermore, when studying particularly rare or elusive species, geneticists are sometimes presented with a lack of biological material altogether. This has been the case with *Mobula rochebrunei*: any studies of the Mobulidae as a group have been forced to rely entirely upon museum specimens for this species, which are usually highly degraded due to their age and preservation, or upon available sequences generated by previous studies.

Despite these problems, the Manta Trust has worked extensively with researchers and conservationists worldwide to compile what it believes to be the most comprehensive set of mobulid genetic samples in the world, both in terms of species, geographical locations, and numbers of individuals. These samples are stored in the UK, and are being utilised by their Global Genetics Project, which uses novel genomic approaches to develop tools for the conservation and management of mobulids. Due to the unclear nature of what constitutes a species within the Mobulidae, and the requirement for effective conservation measures to reflect accurate taxonomy, we are in the process of conducting our own phylogenetic study, using highly advanced DNA sequencing and analysis techniques on multiple samples per species. This publication is in preparation, and the preliminary results contradict some of the latest published findings. We therefore use nomenclature supported by these results in this guide.

The transition from single gene to genome scale projects has been made possible due to substantial reductions in the cost of Next Generation Sequencing (NGS). As a result, we can now use NGS techniques to generate vast quantities of DNA sequence data, far beyond what was attainable for generations of scientists before us. The major limiting factor today is the cost of sophisticated computing facilities capable of storing and processing such large datasets and the requirement for more specialised data analysis.

Nevertheless, when carrying out phylogenetics (the study of relationships between taxonomic groups), or species traceability analyses using genetic or genomic data, specialist knowledge is required, without which data can be misinterpreted, leading to spurious conclusions. For example, it is important to consider factors such as introgression (where genetic information is transferred between species as they hybridise), admixture (where genetic information is exchanged between previously isolated groups, which have begun to hybridise) and incomplete lineage sorting (where the phylogenetic tree as based on any one gene region, or locus, may not in reality correspond to the actual species tree). By doing so, it will be possible to disentangle real biological signals in the data from signals created by any one of these, or many other factors.

Despite these challenges, the study of genetics, and more recently genomics, holds great promise for species identification, which is the ultimate aim of this guide. This can then be applied to the enforcement and monitoring of conservation regulations, such as CITES. What's more, the use of genetic data doesn't stop there. DNA can also help to inform conservation and management initiatives in other ways. These can include identifying the geographical extent of isolated populations, investigating how populations are locally adapted, or quantifying how genetically diverse a population is. These insights can help tailor the management of populations and indicate a population's resilience to external impacts such as increasing fishing pressure, or, pressingly, climate change.

Feeding strategies

When manta and devil rays open their mouths to feed, unfurling those hornlike projections, the cephalic fins, they transform into feeding machines. The once-flattened body and mouth becomes a giant black hole, with the pectoral fins serving as wings that power the ray through the water. Countless litres of plankton-rich water pour in through the mouth and stream out over the five pairs of gill slits that line the throat. Any plankton or small fish larger than a grain of rice is filtered out of the water by feathered gill plates which encircle the gill slits, redirecting the tiny morsels of food back towards the ray's throat. Once the devil ray or manta has trapped a mouthful of planktonic food it closes its mouth and gills, squeezing out the remaining water, before swallowing the highly nutritional prey.

This highly effective feeding mechanism has enabled these rays to exploit one of the oceans' most abundant food sources. But while their prey may be abundant, it is not evenly dispersed throughout the oceans. As a result, manta and devil rays have become experts at locating concentrations of food, seeking out planktonic hotspots that are often ephemeral, forming with the ebb and flow of the tides or the shifting of the seasons.

To maximise their planktonic rewards for the energy they expend while feeding, mobulid rays have evolved a wide variety of feeding strategies. The overall feeding technique is simple. The ray swims through the water with its massive mouth agape, and the paddle-like cephalic (head) fins unfurl in front of the mouth to funnel plankton-rich water through their specially adapted gills. But variations in the rays' swimming positions, and the strategies they use as a group, are key to their feeding success. These strategies can often be complex, requiring coordination among multiple individuals to enhance feeding efficiency. Only recently have scientists begun studying these techniques in depth, but to date, a total of eight different feeding strategies have already been described. However, only half of these feeding strategies (1, 3, 5 & 8) have been observed being undertaken by devil rays. As scientists begin to focus more on the manta's smaller relatives, further feeding strategies are likely to be recorded for these more elusive species.

The zooplankton prey of manta rays and other large filter-feeding animals consists of hundreds of different species which, despite their tiny size, occur in such abundance that they sustain many of the oceans' largest creatures.

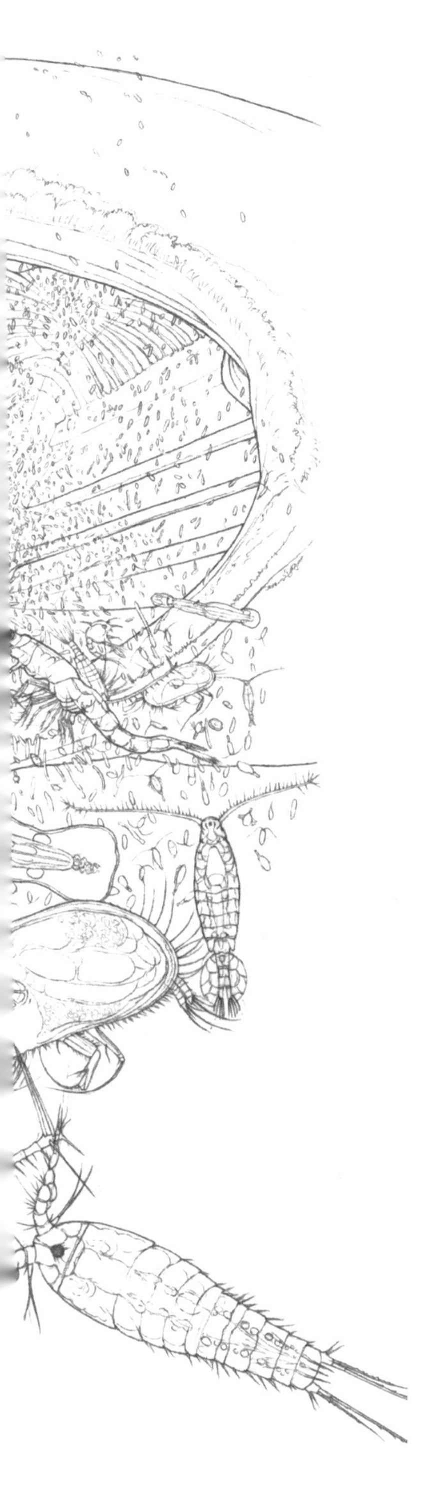

1. Straight feeding (above)

Each mobulid feeds independently, swimming forward in a straight line with its cephalic fins held open in front of the fully open mouth so that they almost touch in the centre. The rays usually feed horizontally through the water; however, sometimes vertical feeding up and down through the water column is observed. The rays perform a sharp 180 degree turn at the end of each 'feeding run', before commencing back along the same plane, feeding in the opposite direction. For manta rays, feeding runs may extend from just a few dozen metres to several hundred metres, depending on the concentration and distribution of their prey.

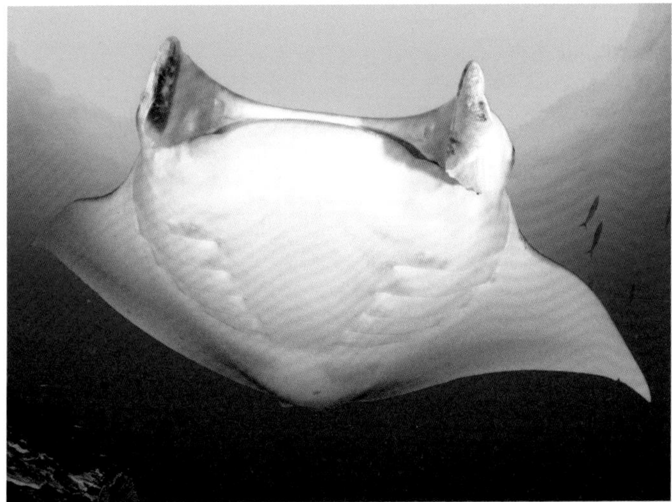

Bottom-feeding manta rays sometimes rub their cephalic fins, gill covers and bellies raw on the seabed while capturing their prey.

2. Surface feeding (right)
Feeding independently, the manta ray positions itself just below the water's surface, tilting its head back so that the upper jaw of its mouth is just above the water. The close proximity to the surface means the manta has to reduce the up-stroke of the pectoral fin to prevent its fins from lifting above the water's surface. The cephalic fins are positioned in front of the mouth in the same manner as straight feeding; although the mouth is usually only three-quarters open. The feeding runs also follow the same pattern as straight feeding, although in order to execute the 180 degree turn at the end of each run, the manta is required to undertake a slight dip away from the water's surface.

3. Chain feeding (above and right)
Lining up head-to-tail, the manta or devil rays form a line of as many as several dozen individuals moving through the water column together along the horizontal plane. Mouth and cephalic fins are held in the same position as straight feeding and at the end of each feeding run the chain of rays often continue to hold the line, which snakes around behind the leading animal. Similar to a flock of birds flying in a 'V' formation, the following rays often position themselves slightly above or below the individual in front of them, like these West Atlantic Pygmy Devil Rays *Mobula hypostoma* pictured above.

4. Piggyback feeding (right)
Feeding together in close proximity, a smaller individual, usually a male manta, positions itself directly on the back of a straight-feeding larger individual, usually a female, matching the beats of its pectoral fins to the beats of the larger individual. Occasionally, several individuals piggyback on top of one manta, resulting in stacked feeding of three, or even four, manta rays, all swimming horizontally through the water column together. At the end of a feeding run when the lowest positioned individual turns to swim back in the opposite direction, the piggybacked individual/s are usually displaced. Piggyback feeding has only been observed in Reef Manta Rays *Mobula alfredi*.

5. Somersault feeding (left)
Feeding individually, the manta or devil ray performs a tight backward somersault in the water. The ray completes a 360 degree loop in the water column, the diameter of which is less than the disc width of the animal's body. Individuals may occasionally rapidly accelerate into a backward somersault, and in these instances usually only one or two complete loops are performed before the ray resumes straight feeding. However, as many as several dozen continuous backward somersaults may be performed before the animal breaks the looping cycle and returns to straight feeding.

6. Cyclone feeding (above)

This behaviour has only been recorded in Reef Manta Rays *Mobula alfredi*. A line of chain-feeding mantas begin to loop around until the lead of the feeding chain joins the trailing mantas to form a large circle of feeding animals. As more and more animals join the circle, the column of mantas builds through the water to resemble an underwater cyclone of mantas which is approximately 15 metres in diameter. The spiralling mass of as many as 150 individual manta rays circle around together for as long as 60 minutes. The rotating cyclone always turns in an anticlockwise direction when viewed from above looking down onto the mantas.

7. Sideways feeding (right)

Only recorded being undertaken by the manta ray species, the behavioural characteristics of sideways feeding are similar to straight feeding, except that during the feeding runs the manta flips itself sideways, rotating the plane of its body 90 degrees away from the normal horizontal feeding position. The cephalic fins are also held out in a position perpendicular to the plane of the body, away from the manta's head; very different to the standard cephalic fin position. The manta makes feeding runs backwards and forwards, similar to chain and straight feeding, returning the plane of its body to the normal horizontal swimming position only during the turns between each feeding run.

8. Bottom feeding (right)
Feeding individually, the manta or devil ray swims along the seabed with its open mouth positioned within a centimetres of the bottom. The seabed forms a natural barrier to the ray's prey, so in manta rays the animal's unfurled cephalic fins are usually splayed apart, positioned out away from the mouth. This funnels any plankton in front of the approaching manta ray in towards the centre of the mouth. In devil rays the cephalic fins are held in their more common central feeding position during bottom feeding.

A ninth feeding strategy, **lunge feeding (below)**, yet to be formally described for mobulid rays, has also been observed by the authors. This feeding behaviour is employed by devil ray species which feed on faster moving prey, like opossum shrimps and schooling fishes; such an anchovies, silversides and lanternfish. The devil rays attack dense schools of prey in groups: accelerating as they approach, they break formation as they pass through the school, scattering their prey, which become easier to catch.

A feeding group of Longhorned Pygmy Devil Rays *Mobula eregoodootenkee* attack a shoal of tropical anchovies *Stolephorus* spp., scattering their prey in all directions as they accelerate through the shoal.

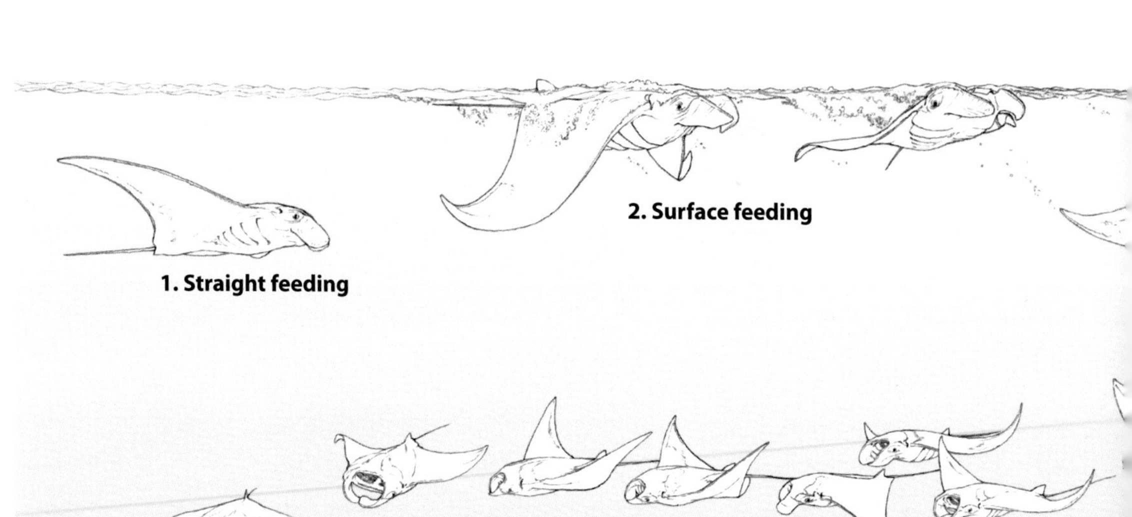

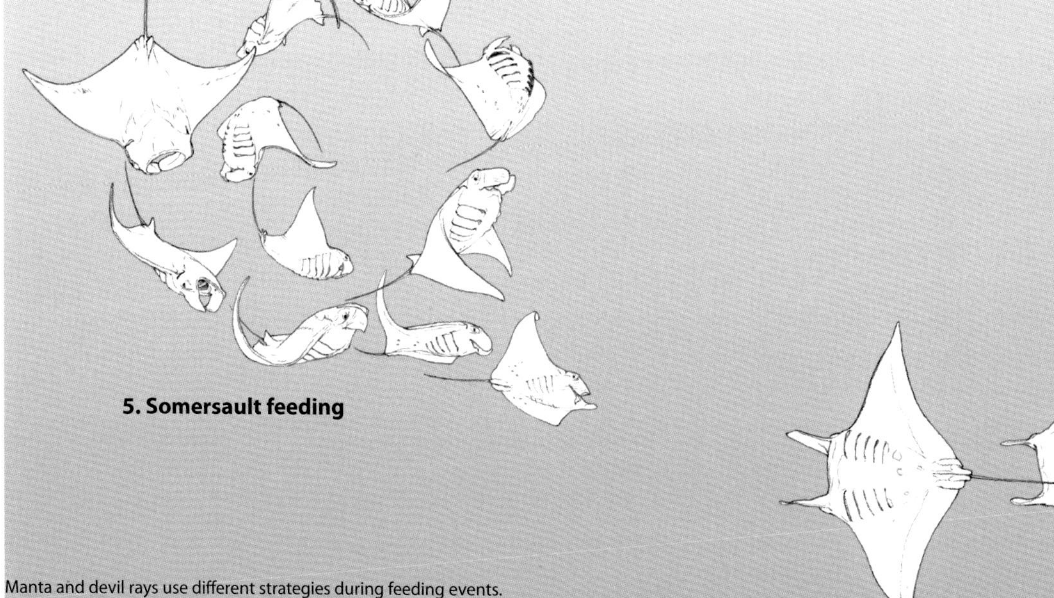

Manta and devil rays use different strategies during feeding events. Some are simple, employed individually – 1. straight feeding, 2. surface feeding, 5. somersault feeding, 7. sideways feeding and 8. bottom feeding. While others involve group behaviour that may see dozens of individuals feeding together to capture their prey – 3. chain feeding, 4. piggyback feeding and 6. cyclone feeding.

6. Cyclone feeding
8. Bottom feeding

Leaping devils

Do manta and devil rays breach? Yes, regularly, and they can do so with surprising agility, at times leaping several metres clear of the water with just one powerful thrust of their wings and landing belly side either up or down. It seems most mobulids around the world visit the other side of the looking glass every now and again, even if it is for just for a second or two. But why do they jump?

Nobody really knows, but there are a few likely explanations. The first is to dislodge parasites and the irritating remora fish that often cover their bodies, while a second could be communication. If communication is driving some of the breaching behaviour observed, it would suggest a much higher degree of social interaction than has previously been recorded. What we do know is that when a tonne of manta ray falls back into the ocean, it creates a massive splash above water and a resulting pressure wave below that could easily carry several miles through the water in every direction. Manta researchers have felt and heard this pressure wave pass through their bodies when a jumping manta landed several metres away from them, so it is entirely possible for other mantas within range to detect the pressure wave as it passes through theirs.

At some manta feeding aggregation sites it is not uncommon to record several dozen manta rays leaping clear of the water within just a few hours. The frequency of these leaping events also seems to concentrate around the start of a feeding session. It is possible that the leaps and the pressure waves they create alert other mantas in the area to the feeding opportunity. But what does a leaping manta gain by advertising this to the other mantas?

To answer this question we have to go back to the feeding strategies of the mantas. Remember that manta rays often rely on group feeding with other

It is thought that breaching behaviour in manta and devil rays may be a way to communicate with each other by using pressure waves that radiate away through the water column from their bodies as they splash back into the sea. Here the Spinetail Devil Ray *Mobula mobular*.

individuals to maximise the food they can consume in a finite period. Indeed, feeding events are often limited not by the time it takes the plankton feeders in the vicinity to consume all of the food, but by the turning of the tides. The animals are not competing against one another but against the natural elements. The race to feed, to gobble up as much plankton as they can, is a race against the outgoing tides that will carry the plankton to sea again. In this scenario, breaching as an intentional form of signalling would lead to increased group feeding, reinforcing the social interactions among individuals and improving the overall feeding success for each manta in the population.

Devil rays also regularly breach, often leaping several metres out of the water in a spectacular display of aerial acrobatics. These breaching events regularly occur during mass aggregations of these smaller relatives of the mantas. In the Sea of Cortez in Mexico these aggregations occur at certain times of the year when thousands of individual Munk's Pygmy Devil Rays *Mobula munkiana* come together, most likely to mate. The breaches occur most often around the outside edges of the shoal of rays – like rockets exploding out of the water, the pygmy devils probably make these leaps as part of a courtship display, advertising their fitness and readiness to mate. As the rays land they slap their pectoral fins down flat on the water's surface, possibly trying to create the loudest splash for all to hear.

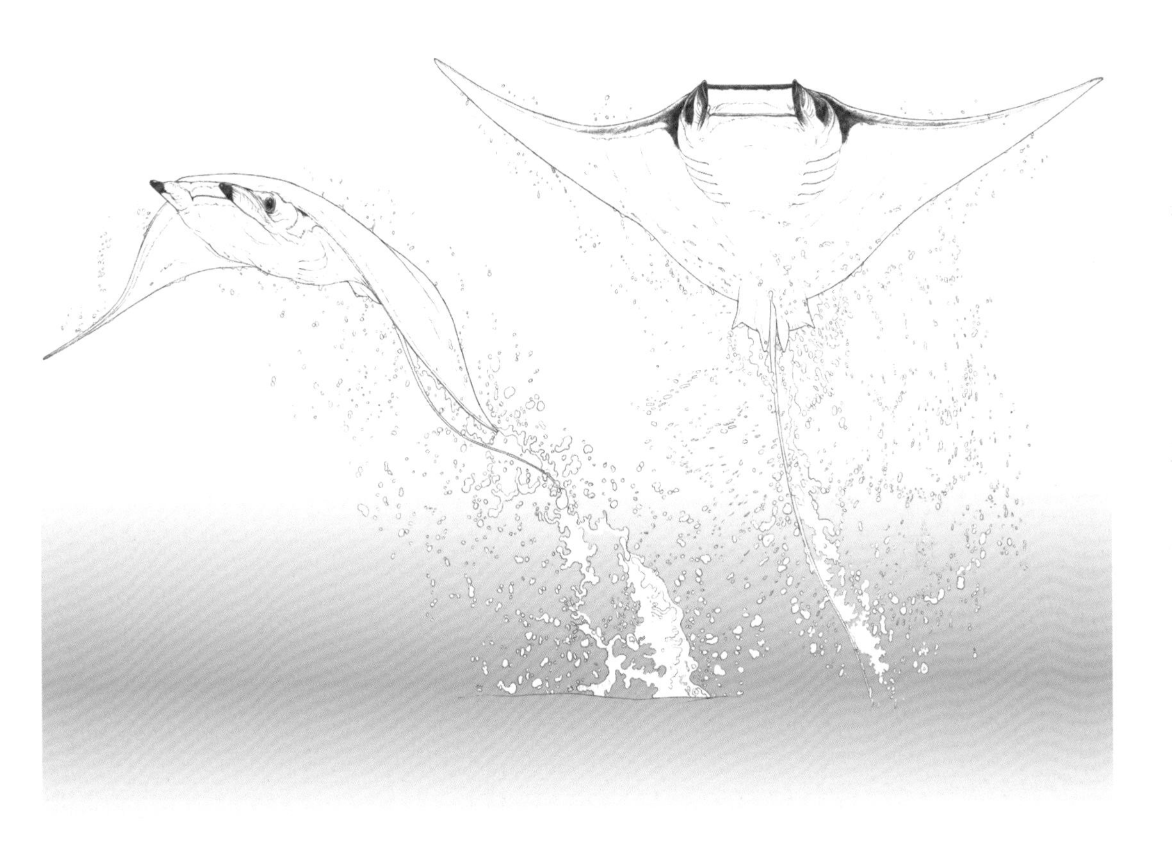

Munk's Pygmy Devil Rays *Mobula munkiana* leap several metres out of the ocean during mass aggregations of these rays, which can number in the tens of thousands.

During courtship events, a female manta ray which is being chased by males also leaps clear of the water, possibly signalling to other potential suitors in the area to come and compete for her amorous attention. For the choosy female, the more males she can attract into the area, the better.

These speculative theories still need a great deal more research, but if mantas and devil rays are indeed communicating with each other using pressure waves, then these animals have a much greater level of social awareness and intelligence than they currently get credit for.

One common myth surrounding breaching manta rays is that the pregnant females breach to give birth. This fishy tale arose from fishermen who observed manta rays giving birth while leaping into the air after becoming ensnared on a hook and line, or after being harpooned. Pregnant females of other shark and ray species are also known to abort foetuses during periods of extreme stress, such as becoming inescapably entangled in the fishing nets or during an attack from a predator.

Aborting the foetus prematurely is a desperate act by the mother as she tries to save either her own life, or

Reef Manta Ray *Mobula alfredi*, at sunset, Baa Atoll, Maldives.

Munk's Pygmy Devil Rays *Mobula munkiana*, as the rays land they slap their pectoral fins down flat on the water's surface, possibly trying to create the loudest possible splash for all to hear.

the life of her unborn pup/s. Under natural predatory circumstances it is possible for the mother to save her own life by distracting the attacker with an easy meal in the form of her aborted foetus. If the attacking predator ignores the foetus and has already inflicted mortal injuries on the mother, by aborting her pup/s there is a small chance that her young might be able to survive, despite being born prematurely.

Right: a pair of Spinetail Devil Rays *Mobula mobular* raise their heads and cephalic fins above the water during a courtship event, during which the male attempts to position himself on top of the female. Below: Munk's Pygmy Devil Rays *Mobula munkiana*.

Cleaning giants

The world's oceans can be an itchy place for their inhabitants to live, especially the larger ones, whose bodies present a much bigger target for waterborne parasites. Ectoparasites live and feed on their hosts' bodies, hiding themselves inside mouths, spiracles, gills and anywhere else they can secure a firm grip out of harm's way. Filter-feeding animals like the mobulid rays, which spend much of their time swimming around with their mouths wide open, are also susceptible to particles of food and other debris becoming lodged in their gills, creating an unwanted build-up of detritus.

The natural world is full of parasites. Even the parasites have parasites, and the larger you are, the more space and micro-habitats you provide for these tiny creatures to live on and in. On land, larger animals such as birds and mammals often cope with this problem by grooming and cleaning themselves and other members of their social group, using teeth, hands, beaks and claws to rid themselves of irritating parasites. But fish like manta and devil rays do not have hands or beaks, so they have found an animal group to do the job for them.

These animals are called cleaners, comprising mostly small reef fishes, although many shrimp species have evolved to become specialised cleaners as well. The cleaners are all small from necessity. Their size enables them to survive on and pick off the individual parasites that cover their clients, even venturing right inside the bodies of their hosts to reach areas that would be inaccessible to a larger cleaner. But their small size makes the cleaners less mobile, so instead of travelling around in search of clients to clean, their clients come to them. The cleaners set up shop at specific locations, usually a prominent reef outcrop or a coral head, and a whole host of client species make the trip each day to visit these 'cleaning stations'. All manta ray species and the

Shorthorned Pygmy Devil Ray *Mobula kuhlii* have been recorded frequenting cleaning stations, and it is likely that further investigation will also reveal some of the other devil ray species utilising these cleaning stations.

This relationship between cleaner and client has long fascinated marine ecologists. It is widely showcased as the perfect example of mutually beneficial symbiosis, with both cleaner and client receiving an overall benefit from this cooperative behaviour. For the cleaners, the benefit comes from a ready supply of food brought to them on a plate – in this case, upon the body of their clients. The cleaners remove irritating ectoparasites and algal growth, they clean wounds by picking off dead skin and infectious bacteria and fungus, and they perform dentistry by swimming inside the mouths of their clients to remove food particles that have become lodged inside.

For the clients the benefits seem obvious – having a local dentist, doctor and body salon all rolled into one package, with limited waiting times and no reservations required. But the more scientists study these symbiotic relationships, the more complex and, in some cases, less mutualistic they appear to be. A few experiments have suggested that some cleaners engage, at least occasionally, in behavioural parasitism, exploiting the sensory system of their clients by using tactile stimulation to pacify their clients so they can feed on their body mucus, or even their flesh. Overall, however, the relationship is still thought to be beneficial to both parties, with true mutualism at the root of these behavioural interactions. Certainly, the large extent to which cleaning partnerships occur throughout the marine environment, both geographically and taxonomically, would seem to support this theory.

Say ahh! A Reef Manta Ray *Mobula alfredi* hovers above a cleaning station (opposite and above), opening its mouth wide to allow cleaner fish Blunthead Wrasse *Thalassoma amblycephalum,* Lyretail Wrasse *T. lunare* and Blue-streak Cleaner Wrasse *Labroides dimidiatus* to search right inside for parasites.

The ocean is an itchy place, especially for giants like manta rays, which attract a host of unwanted parasites that attach themselves to the manta's body. Top: an Oceanic Manta Ray *Mobula birostris* hovers above an undersea ridge extending out from the island of Socorro in the remote Revillagigedo Archipelago as Clarion Angelfish *Holacanthus clarionensis* flock to its back, grazing on the parasites and thick mucus layer borne on the giant black dinner plate. Below: a Reef Manta Ray *Mobula alfredi* in the Maldives scratches itself on the sandy seabed next to the cleaning station in Hanifaru Bay.

Predators and wound healing

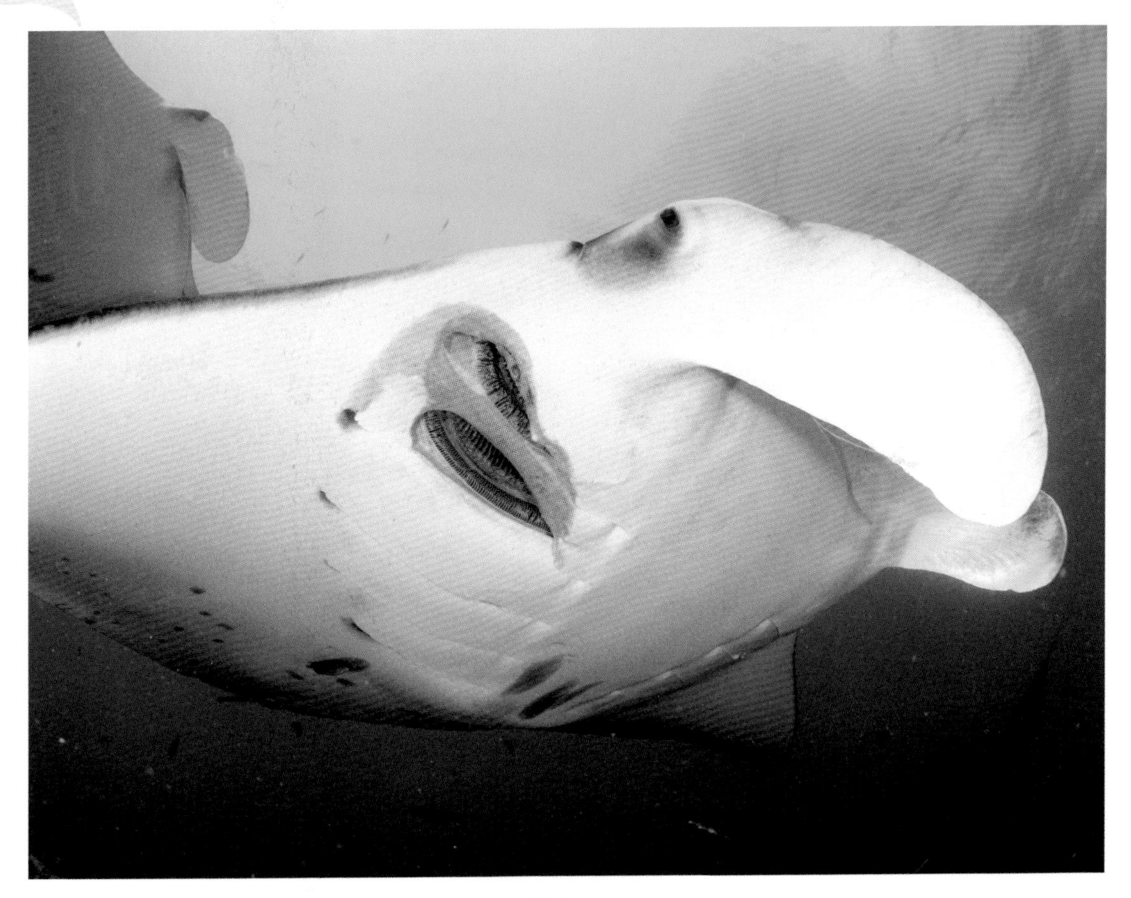

Because of their large size, adult mobulids have few natural predators, with only the oceans' largest predatory sharks and marine mammals able to kill a fully grown adult; such as the Tiger *Galeocerdo cuvier*, Great Hammerhead *Sphyrna mokarran* and Bull Sharks *Carcharhinus leucas*, or the False Killer Whales *Pseudorca crassidens* and Orcas *Orcinus orca*. This is especially true for the manta rays and the three larger devil ray species, which often escape with only a bite on the trailing edge of one of their pectoral fins. Little data exists for devil rays; however, predation rates on manta rays appear to vary considerably in different locations

The injury to the gills of this Reef Manta Ray *Mobula alfredi* (top) are either the result of a gill infection, or a predatory attack by a cetacean or large shark, like the Tiger Shark *Galeocerdo cuvier* pictured above.

around the world. In the Maldives manta rays have few natural threats to worry about. With just a dozen new predatory attack bites and scars recorded in the study population each year, less than 15% of this population show any evidence of natural predatory attacks. Mortality rates in this population appear to be very low and virtually all of the mantas that have been observed with fresh wounds are sighted again months or years later, their injuries having healed.

Flesh wounds that remove portions of the posterior section of the pectoral fins can also recover remarkably quickly, with large sections of missing flesh able to regenerate with scar tissue to fill in the missing portion of wing over the years. But severe bites that cut deep into the manta's body, removing whole sections or ends of the pectoral fins, will never allow regeneration of cartilaginous tissue in these areas, leaving the mantas with major disfigurements.

In Mozambique the resident population of reef manta rays suffer from much higher levels of shark attacks than in the Maldives. Virtually all of the studied adult population show some form of injury and scarring from predatory attacks. In both of these locations an injured manta ray will spend significantly more time visiting a cleaning station to have its wounds cleaned. In Mozambique small butterflyfishes (Sunburst Butterflyfish *Chaetodon kleinii*) act as the wound specialists, removing dead and infected flesh, helping the injured area to heal faster. In the Maldives the Blue-streak Cleaner Wrasse *Labroides dimidiatus* performs this task.

Manta rays have an amazing ability to survive serious injuries, like the two individuals on the opposite page. The male Reef Manta Ray *Mobula alfredi* to the right was photographed within a few weeks of sustaining a shark bite which removed a large section of his right pectoral fin (top). Twelve months later scar tissue had regenerated much of the pectoral fin area removed by the bite (below).

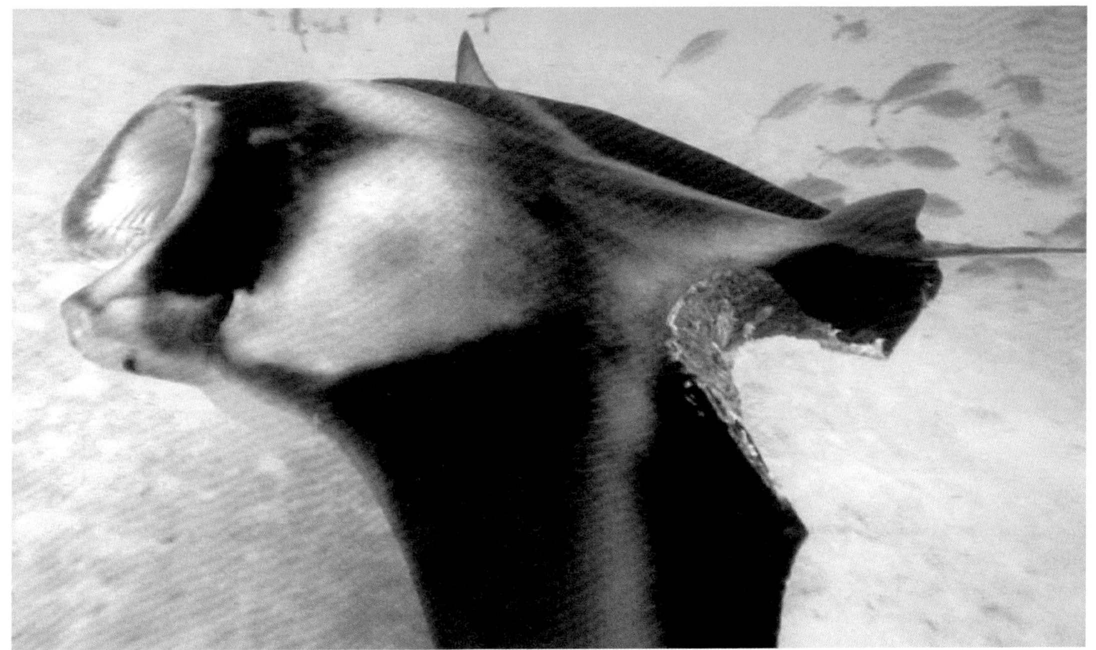

Intestinal eversion and vomiting

Filter feeding with your gills inevitably means you're going to get some unwanted clogging. Manta rays deal with this problem by coughing, often at cleaning stations, which back-flushes water through the fine weave of their gill plate lobes, dislodging trapped food particles and parasites at the same time. The resultant cloud of coughed up gill detritus drives the cleaner fish wild, and they rush to gorge themselves on the food bonanza. Some individuals even appear to vomit large clumps of the undigested exoskeletons of their zooplankton prey.

A similar event occurs when manta rays defecate, which they also often do at cleaning stations. This results in a cloudy stream of dark red plankton defecation sometimes wafting across a group of divers who are often too slow to react.

Manta rays can also evert their intestines as far as 30 centimetres (12 inches) from the opening of their cloaca, essentially squeezing part of their intestines outside their bodies. They often do this above cleaning stations, repeating the eversion process four or five times in rapid succession and defecating at the same time. This is likely a good way for the mantas to clean their intestines and dislodge internal parasites. Because manta rays' excrement is often a dark red colour, and because the intestines come so far outside the animal's body, divers often mistake defecation and intestinal eversion for manta rays they believe are bleeding or giving birth.

Devil rays have not been recorded vomiting or everting their intestines, but this is probably because they are much harder to observe in their natural habitats than mantas.

Manta rays defecate regularly (top), especially at cleaning stations, which provides a bonus meal for the hungry cleaner fish. To dislodge intestinal parasites and clean out their stomachs, manta rays also evert their intestines (middle) and vomit the undigested exoskeletons of their prey through their mouths (bottom). Because manta rays' excrement is often a dark red colour, divers often mistake defecation for bleeding.

Remoras and hitchhikers

Manta and devil rays are like floating islands; oasis sanctuaries for a whole host of animals that seek the shelter, protection and sustenance these giants can provide. Tiny copepod parasites attach themselves to every surface of their bodies – inside their gills, around their mouths and in the mucus on the manta rays' backs. Juvenile Golden Trevalleys *Gnathanodon speciosus* ride the pressure waves in front of the giant's head (page 39, top); like dolphins bow riding a ship. These tiny companions appear to lead the mantas, piloting them on their voyage across the open seas. As soon as they are large enough to survive by themselves, the Golden Trevalleys leave the shelter of the rays to seek protection together in shoals.

Another group of fishes have taken their association with, and dependency on, their larger hosts to a whole new level. These fish are called remoras or suckerfish. As the name 'suckerfish' suggests, they suck on to the bodies of their hosts using a modified dorsal fin that acts as a suction cup. This enables them to cling onto the bodies of the manta rays, saving energy. They even venture deep inside their hosts' mouths to rest, find food or escape predation. One species, the White Remora *Remora albescens*, may spend its whole adult life inside the mouth of a manta ray, which must be a little discomforting for the host, especially when the remora erects a spiny fin to prevent it from being squashed when the manta closes its mouth.

The benefits to the remoras and suckerfish of living in such close association with their larger hosts are clear, but there is debate about whether the manta rays receive a significant return in benefits from these relationships. Certainly, there are several negative impacts, such as reduced swimming efficiency through extra drag created by having to carry hitchhikers around, skin abrasions and sores created by the constant suction attachments of the remoras, and the overall discomfort of having to put up with as many as two dozen of these fish scurrying around on and inside their bodies. In extreme cases, a Slender Suckerfish *Echeneis naucrates*, which can grow to 50 centimetres (20 inches) in length, may force itself through the narrow gill slits of its mobulid host. Over time this repeated invasive assault causes disfiguring injuries to the ray's gill. Even more disconcerting are observations of remoras taking shelter inside the cloacal opening of

Large Brown Remoras *Remora remora*, which can also be white in colour (top), rarely stray far from the shelter of their hosts, unlike their close relative the Cobia *Rachycentron canadum*, which only loosely associates with mobulids (above).

a manta or large devil ray, with just the remora's tail left poking outside.

These observations have led some scientists to label the remoras and their relatives as parasites, exploiting the shelter their hosts provide while giving nothing in return. However, recent observations suggest that the remoras act, in part, as permanent live-in cleaners, removing some parasites from the bodies of their hosts, thus providing a benefit to the rays. This may be especially true for the Oceanic Mantas *Mobula birostris* and Sicklefin Devil Rays *Mobula tarapacana*, which spend much more time out in the open ocean, infrequently visiting cleaning stations.

Throughout most of the Reef Manta Ray's range, by far the most common species of remora that associates with this species is the Slender Suckerfish. Sometimes as many as 25 of these remoras can swarm around a single manta, although just one or two individuals per manta is more common. The species that associates most often with Oceanic Mantas and Sicklefin Devil Rays is the Brown Remora *Remora remora*, a large, brown, compact remora that prefers life in the open ocean. These remoras rarely stray far from the shelter of their hosts, while the Slender Suckerfish are much more mobile, swimming free from the mantas for much of the time. On occasion, groups of these Slender Suckerfish can be found resting on the seabed, holding out their fins as if signalling to hitch a ride.

The Slender Suckerfish are an inshore species and do not venture into the ocean depths. Therefore, when mobulid rays, or any other host they may be attached to, venture into deeper water, diving to depths of 100 metres (330 feet) or more, the Slender Suckerfish are forced to abandon their host and return to the shallow reefs until they can find another. By contrast, Brown and White Remoras belong to a different genus and appear to stay with their hosts permanently, wherever they go. This diving limitation is likely the reason why the number of associated remoras with Reef Manta Rays varies so much over space and time. At sites where the mantas are regularly making deep dives to feed, the suckerfish associations are likely to be lower. The longer a Reef Manta spends in the shallows on the reef, the more likely it is to pick up hitchhikers.

Between different cleaning and feeding sites in the Maldives there are large variations in the average number of Slender Suckerfish associated with the mantas. Large female mantas also attract more suckerfish, especially those individuals that are heavily pregnant. This is expected, since females in other elasmobranch species have been shown to stop feeding close to parturition and move to birthing grounds. The heavily pregnant females spend more time in the shallow reef areas away from predators and in the warmer waters of the shallows, which may help to speed up gestation. It may therefore be possible to equate the number of Slender Suckerfish associated with any particular manta with the amount of time it has spent since its last deep dive. Increased numbers of Slender Suckerfish on near-term pregnant mantas may also relate to the habit of these fish feeding on the afterbirth of elasmobranchs, as has been shown with Lemon Sharks *Negaprion brevirostris*.

Slender Suckerfish *Echeneis naucrates*, also commonly known as Sharksucker Remoras (above), are common companions for Reef Manta Rays. The suckerfish have a specially modified dorsal fin they can use to stick themselves on to the manta's skin, hitching a free ride. Repeatedly holding on to the manta in the same location often causes red sores and scarring, seen here on the back of this Sicklefin Devil Ray *Mobula tarapacana* (right bottom). Three juvenile Golden Trevalleys *Gnathanodon speciosus* seek shelter close to a Reef Manta Ray *Mobula alfredi* in the Maldives. As soon as the fish grow a few inches longer they leave the relative safety of the manta to form their own schools (right top).

Courtship and mating

Like all elasmobranchs, but unlike the majority of other fish in the sea, mobulids reproduce through internal fertilisation; male and female must physically come together to mate. Manta and devil rays give birth to live young that are small versions of their parents. Ready to fend for themselves, they are completely independent from birth.

Courtship rituals and mating events are observed relatively infrequently, and the only documented accounts of a manta ray giving birth come from a single female housed in a public aquarium in Japan. No devil rays have ever been recorded giving birth under natural circumstances. Many gaps still remain in our knowledge of the life history strategies and reproductive behaviour of these animals, and what little we do know is based on limited scientific data, much of which comes from observations primarily of the Reef Manta Ray. The generalised courtship and mating behaviour described below is, therefore, likely to vary slightly between the different species as more observational data across the family is recorded. It nonetheless serves as a good overview of the reproductive behaviours exhibited by this family of rays.

A male manta ray will often try his luck, testing the female's receptiveness to his amorous advances by shadowing her movements. Positioning himself directly on top of her back, he uses his unfurled cephalic fins to rub the top of her head. Most of the time these enthusiastic advances are rejected; the female literally gives the male the cold shoulder, bucking and twisting her body upwards and away from the male to dislodge him from her back. Sometimes, though, the female responds by rapidly accelerating forward, leading the male around the reef at high speed. This behaviour soon attracts the attention of more amorous males, which follow in line behind the female to form a manta conga dance – the 'courtship train' has begun!

As many as 30 males line up head-to-tail behind the female as she swims at great speeds, twisting and turning around the reef, even leaping from the water. These courtship trains are amazing to watch as the procession of rays snakes through the water, often rushing past like a freight train within inches of divers' heads as the female tries to shake loose the less persistent of her pursuers.

Sometimes the chased female loops back on herself and starts to chase the trailing tail of the courtship train, while at other times she follows other females who get caught up in the action. When multiple receptive females are present, courtship trains can merge or divide. They are in a constant state of flux as new individuals join the queue and old males trail off the back. The trains run hot and cold, reaching periods of explosive speed and energy before slowing again into a graceful procession led by the female.

To watch these underwater ballets is spectacular, but it is not just the sheer exuberance of the chase that drives this behaviour. During the courtship train the female is not really trying to escape her pursuers. Instead, the train is a way for her to assess her suitors; she is testing the males before she makes her choice. This mate selection strategy is primarily driven by offspring viability. After mating, the male ray does not invest any further energy in the future survival of his offspring, leaving him free to pursue other mating opportunities. The female invests heavily, undergoing a long gestation before giving birth. The male's only contribution to the next generation is his sperm, so the female needs to be choosy about her mate to make sure her pup has the best possible chance of survival. The multiple matings the females may undergo during the mating period may also be a way for them to

The enthusiastic advances of a male Reef Manta *Mobula alfredi* are rejected. The female literally gives him the cold shoulder!

ensure a good supply of sperm, which may be stored for a year or more before the embryo is fertilised.

By the time the actual coupling takes place, the courtship train has often tailed off so that just one male remains tenaciously sticking to the female's tail, waiting for his chance. Once the female is ready she slows her swimming speed dramatically, often rising up into the water column as she waits for the male to make his move. Positioning himself on the female's back, he begins to slide his mouth down the top of her left pectoral fin, using his cephalic fins to guide him until the female's wing-tip enters his mouth. The male bites down hard on the end of the female's wing-tip, working as much as one metre (three feet) of her pectoral fin deep into his mouth. Once the male has a good grip on the female's pectoral fin he flips his body around and underneath hers so that they are positioned belly to belly. The male then inserts one of his two claspers into the female's cloaca and ejaculates his sperm with thrusts of his abdominal region. Copulation itself lasts for just 30 seconds and the female remains motionless throughout, while the male continues to beat his pectoral fins, causing the mating couple to spiral around in slow motion in the water column.

All mobulids are negatively buoyant, so during copulation the entwined pair begins to fall towards the seabed as soon as they stop swimming. This is likely the reason why the females usually choose to swim up towards the water's surface before allowing the male to mate. It is also probably why, on most occasions, the females wait until there is just one remaining suitor before they allow mating to occur. Crashing into the reef is not a pleasant way to end a romantic encounter and it can result in injuries.

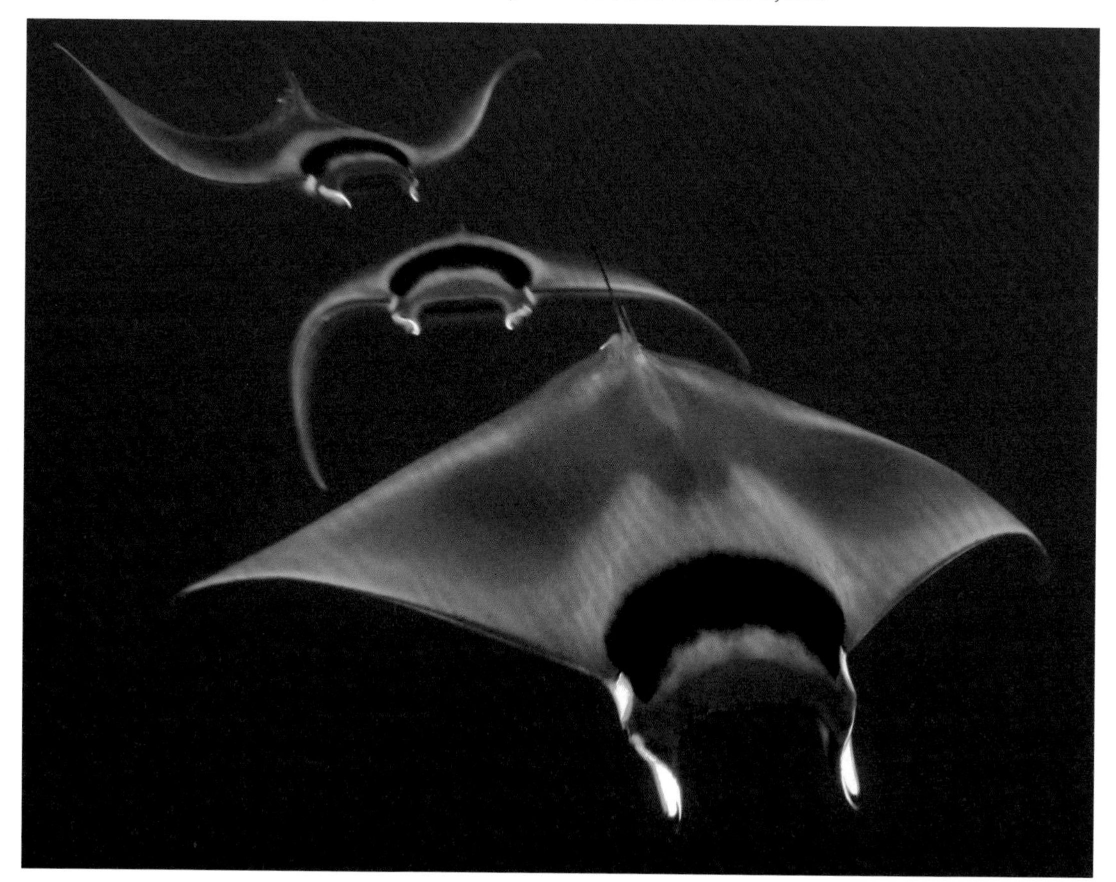

Giving birth to just a single pup, and with no further investment by the males towards their offspring after mating, female manta (page 40) and devil rays (like these Spinetail Devil Rays *Mobula mobular* above), test the fitness of their would-be mates by leading them on a procession, known as a courtship train.

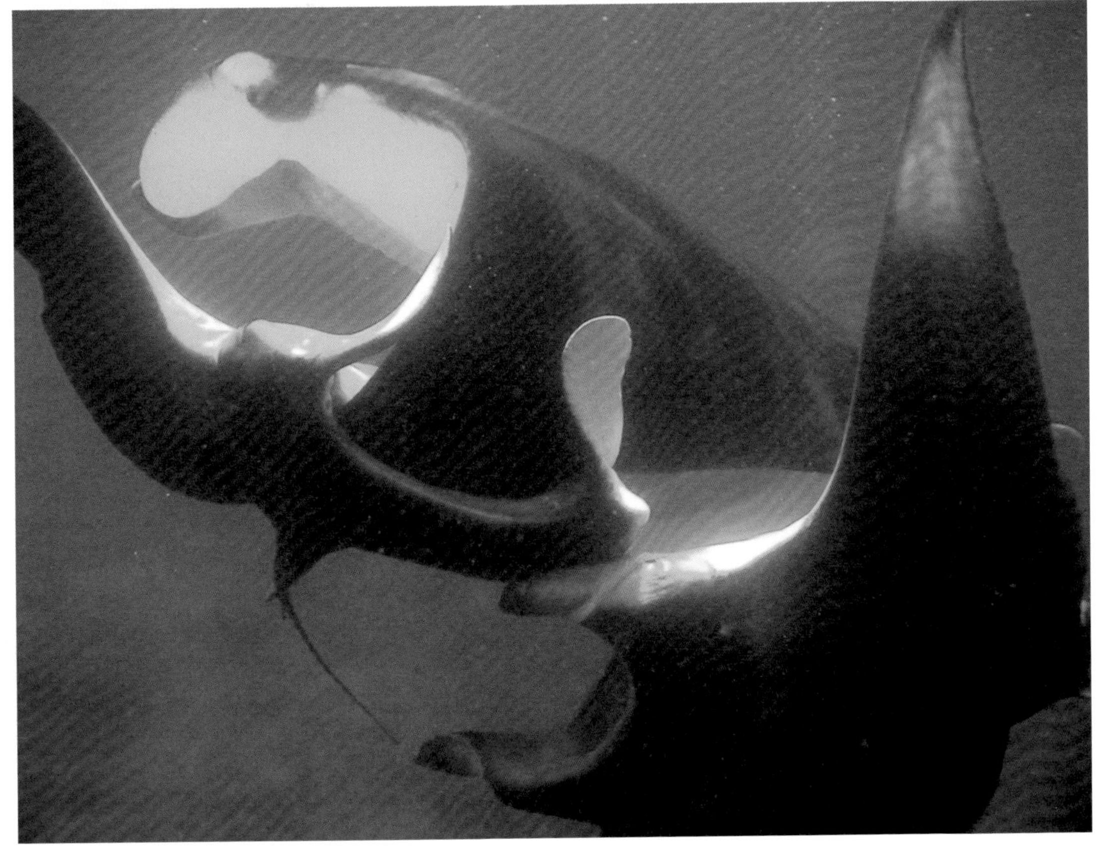

Before mating, male manta rays position themselves on the back of the female (top). Two sexually charged males compete to mate with a female. One male engulfs the female's left pectoral fin in his mouth, biting down hard as he flips and rotates his body underneath the female to insert one of his two claspers (sexual organs) into the female (bottom).

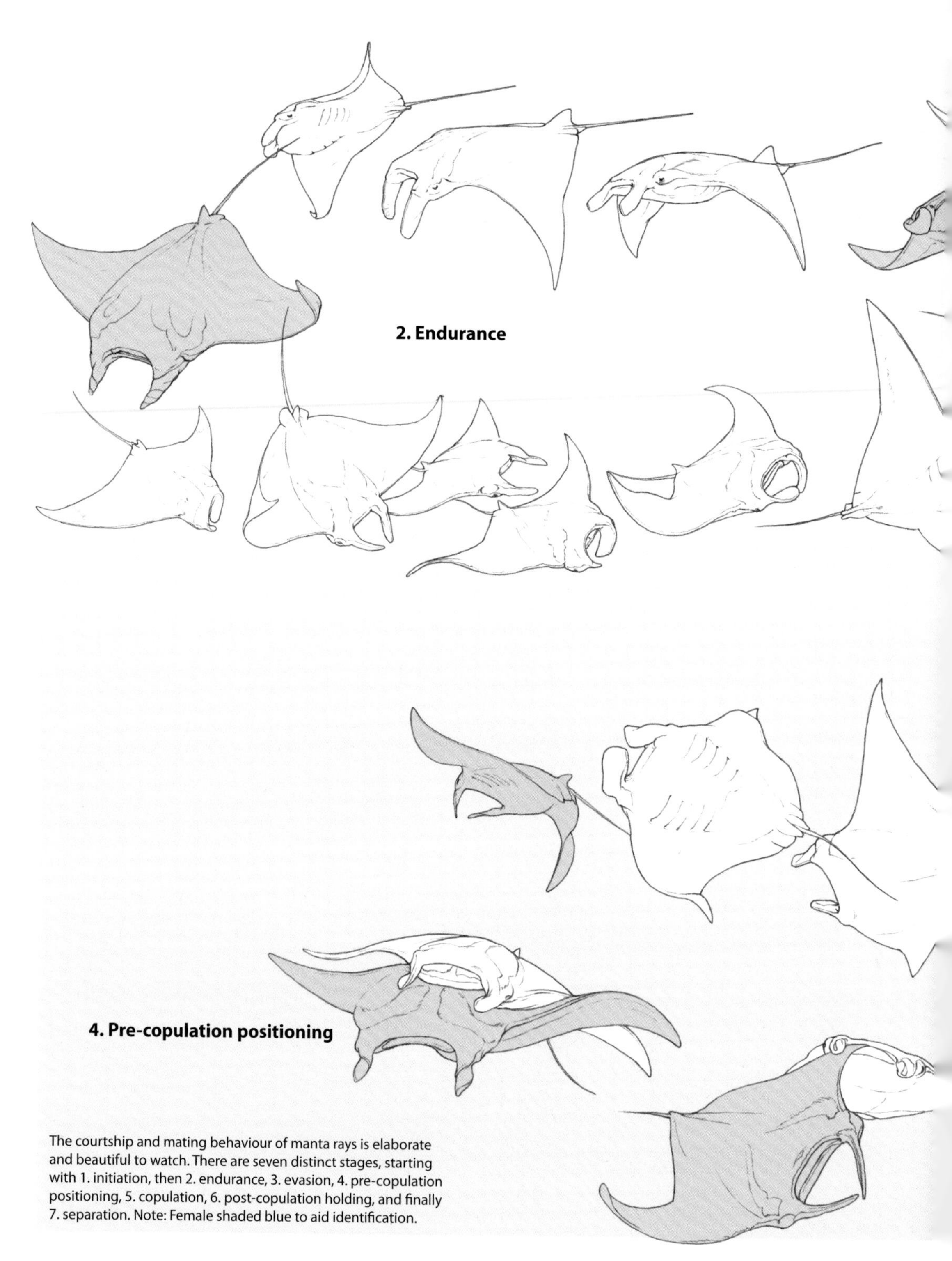

The courtship and mating behaviour of manta rays is elaborate and beautiful to watch. There are seven distinct stages, starting with 1. initiation, then 2. endurance, 3. evasion, 4. pre-copulation positioning, 5. copulation, 6. post-copulation holding, and finally 7. separation. Note: Female shaded blue to aid identification.

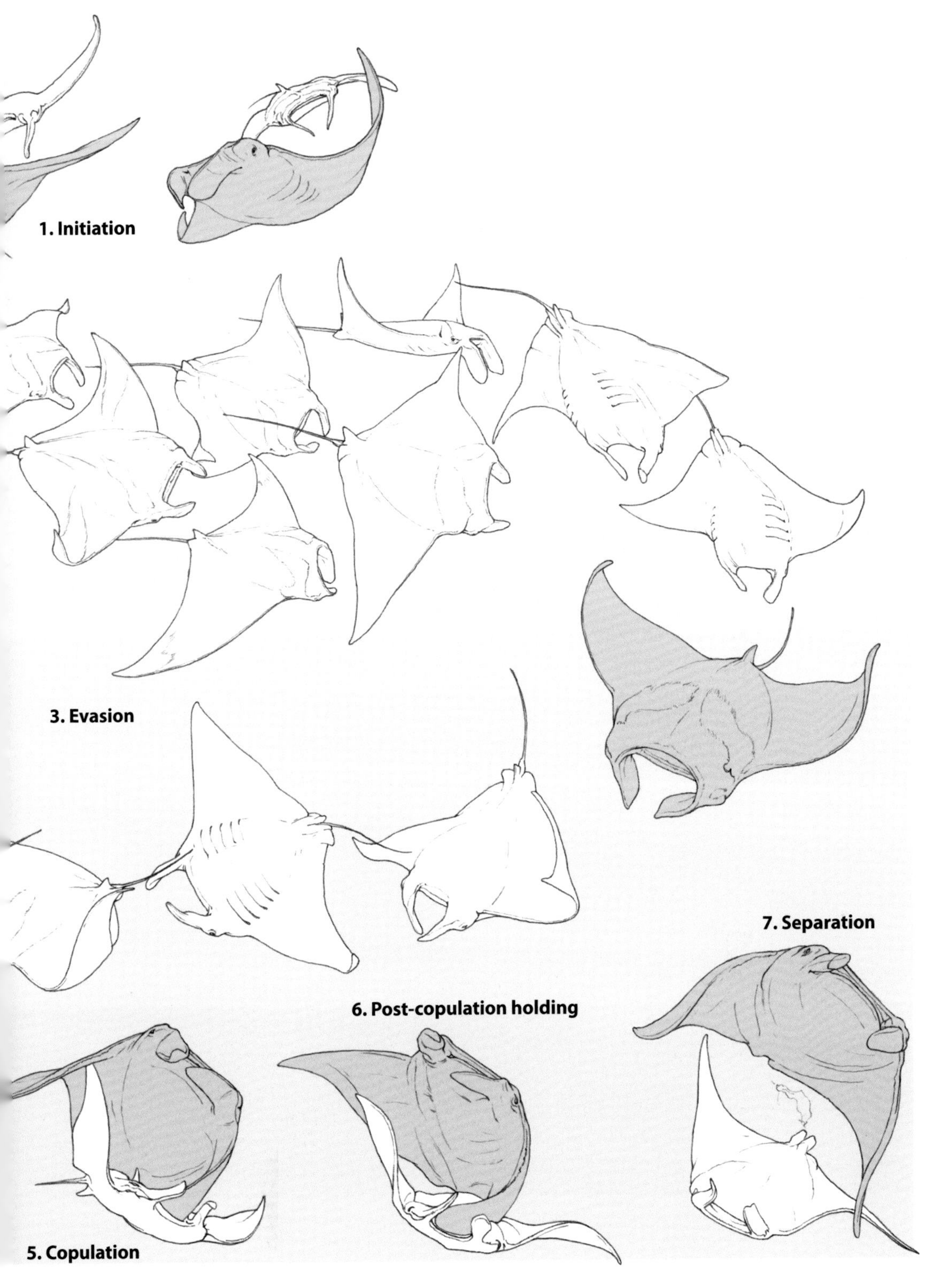
1. Initiation
3. Evasion
7. Separation
6. Post-copulation holding
5. Copulation

Mating scars

When a male manta or devil bites hold of the female's pectoral fin during mating it causes minor cuts and abrasions to the upper and lower surface of the female's wing-tip. The male's teeth on the lower jaw often leave linear scrapes on the underside of the female's fins, which are visible as red scratches. These marks quickly fade against the white coloration of the fin's underside and are often hard to see just a few weeks after mating has occurred.

On manta rays, the male's grip also causes scarring on the dorsal tips of the female's pectoral fins, leaving behind white and/or black circular and oblong marks as the upper layers of skin on the female's fin are scraped away by sharp ridges on the supporting branchial arches in the roof of the male's mouth. These dorsal mating scars are permanent, and while not every mating results in scars, after a female has been through a few matings, most appear to have some permanent visible mark as a memento of their sexual exploits. These love bites allow manta scientists to gauge what percentage of the female population consists of fully mature females. The fresh mating wounds on the underside of the pectoral fins also allow scientists to plot recent matings, which can be used to track pregnancies and reproductive trends over the following years.

Magnified image of the tooth band in the lower jaw of manta rays.

Another interesting observation of these mating scars is their distribution. Over 95% of mating scars are present only on the female's left pectoral fin. This trend for lateralisation is found in both species of manta rays around the world, and is also observed in a wide variety of other species, including humans, most of whom favour the right-hand side of our bodies.

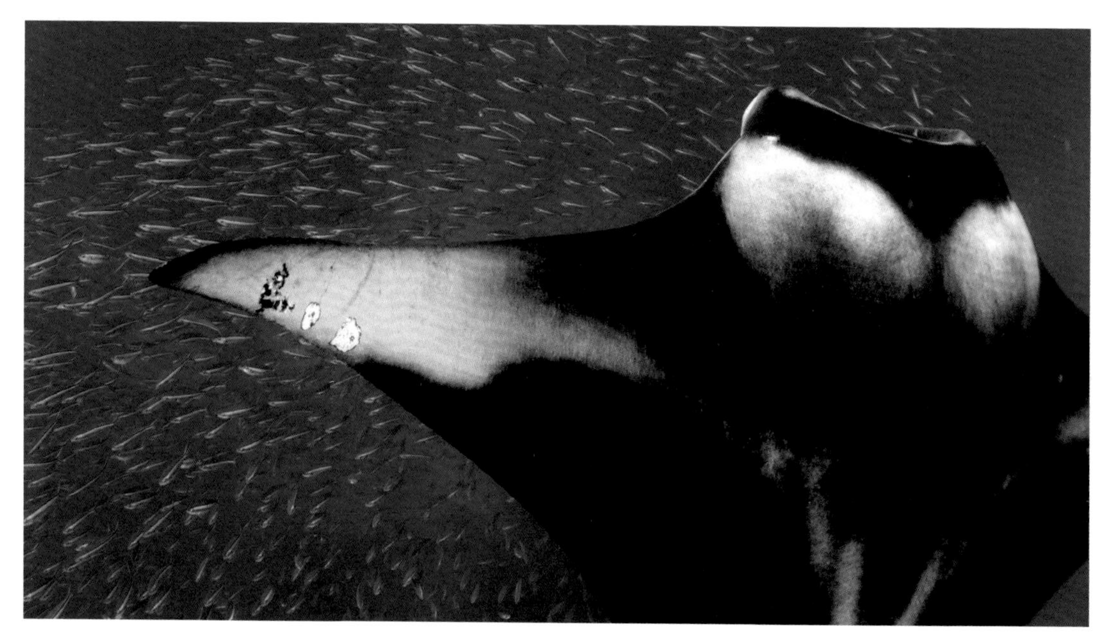

During mating the male manta ray inflicts injuries on the dorsal tip of the female's pectoral fin (above) with ridged cartilage spikes in the top of his mouth (opposite left). These permanent love bites, known as mating scars, help scientists identify which females in a population have mated before.

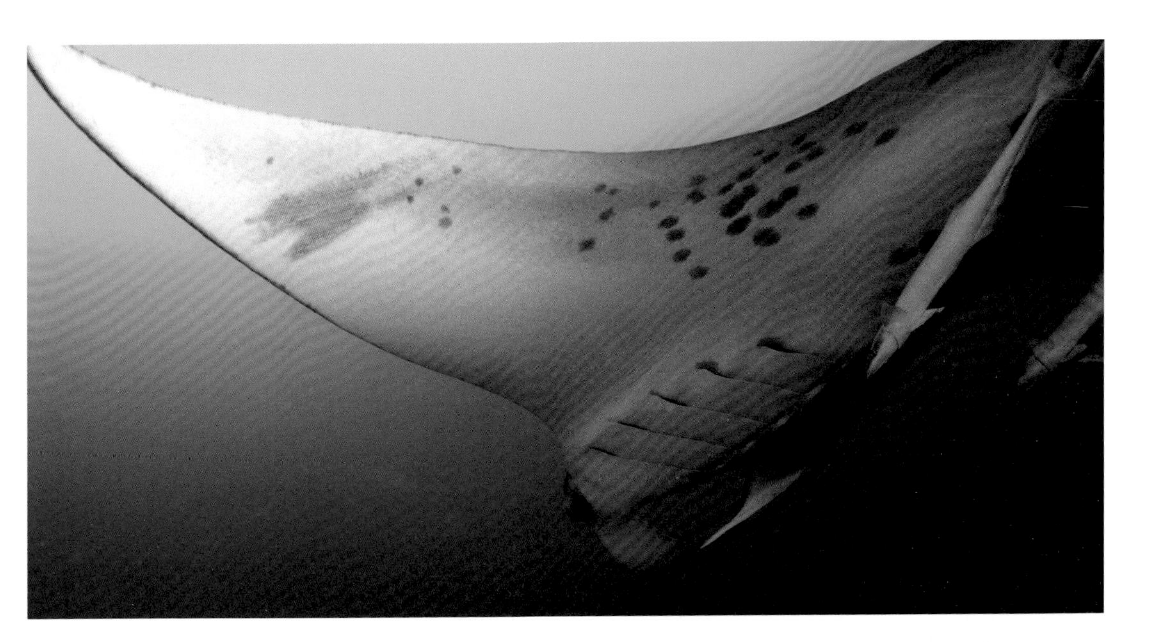

Mating also causes injuries to the ventral tip of the female's left pectoral fin (above). These scratches (red when freshly inflicted) are caused by the rows of teeth in the male's bottom jaw (opposite and right) and can be seen in both manta and devil rays (bottom right).

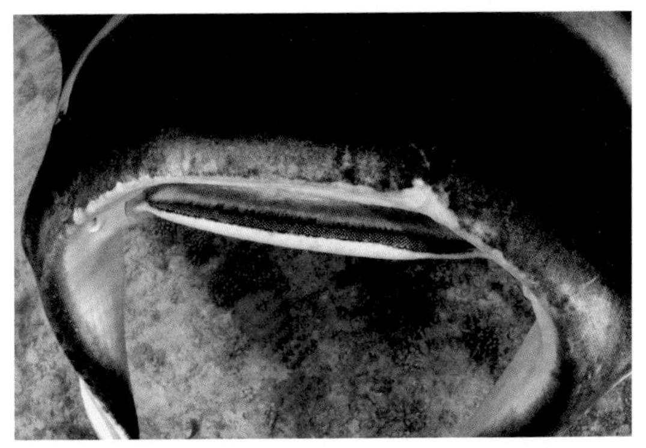

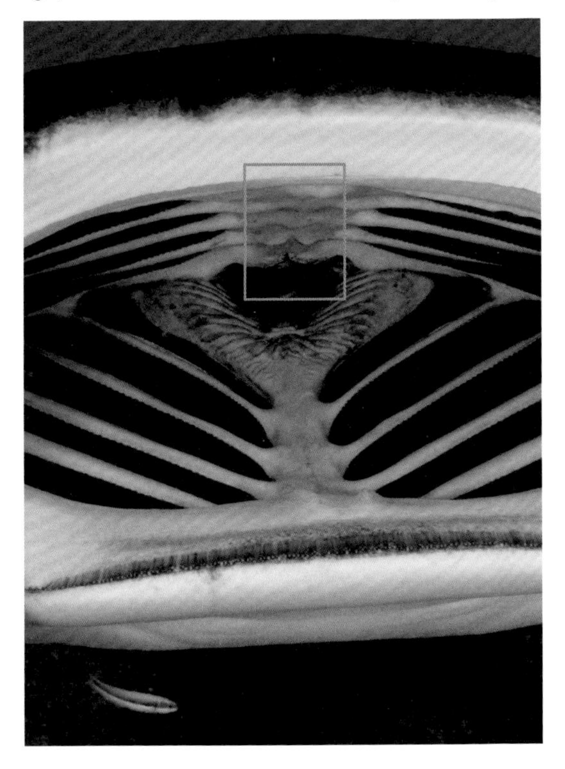

Gestation and pupping

All mobulid rays reproduce via aplacental viviparity, meaning they give birth to live young that are hatched from an egg inside the female's uterus. The pup, which is wrapped in a thin, membranous eggcase, hatches inside the mother's uterus and then feeds on the mother's uterine milk until it is fully developed and ready to be born. No one has ever observed a manta or devil ray giving birth naturally in the wild; however, captive births of manta rays show that the pups pop out of their mother's cloacal opening with their pectoral fins rolled up over their back.

After a gestation period of just over a year, the female Reef Manta Ray gives birth to a single pup (although occasionally they may also give birth to twins) which measures on average 1.5–1.8 metres (5–6 feet) across from wing-tip to wing-tip at birth. The gestation period and pup size at birth for each of the devil ray species and the Oceanic Manta Ray is less defined (although a birth size of >1.6 metres has been recorded for the Spinetail Devil Ray *Mobula mobular*), however each of these species also gives birth to one well-developed, relatively large pup after a long gestation period.

Many parallels can be drawn between the life history strategies of devil rays and some large marine mammals. Unlike their warm-blooded cousins, however, female manta and mobulid rays do not exhibit any parental care for their offspring once they have been born. Accurate gestation periods have been obtained from one female manta ray held in the Okinawa Churaumi Aquarium in Japan, but a lack of consistent and continuous observations of manta and devil rays in all but a few wild populations have made it extremely difficult to record accurate mating and birthing intervals of wild populations.

Under natural conditions, where food is limited and variable, female Reef Manta Rays cannot usually sustain continual reproductive cycles of mating, pregnancies and births without rest periods in between. Even when food is abundant, consecutive pregnancies are hard to sustain due to the high costs associated with producing such a large offspring. The females therefore usually need seasonal gaps in their reproductive cycles to build up their energy reserves. On average, Reef Manta Rays around the world give birth approximately once every two to three years and in some locations the reproductive rate is as low as one pup every seven years.

All manta and devil rays give birth to live young; usually a single, well-developed pup. A female which is close to giving birth clearly shows the bulge of the pup inside her abdomen (above). At birth, a Reef Manta Ray *Mobula alfredi* pup measures on average 1.5 metres (5 feet) in disc width (right).

Big and brainy

Manta and devil rays have brains that are disproportionately large when compared to their body weight. The weight of a manta's brain is more comparable to that of a similar-sized mammal – much larger than the brains of similar-sized fish.

Not only do these rays have the biggest brains of all fish, but the regions of the brain that account for this enlargement are the telencephalon (the anteriormost area of the brain) and the cerebellum (the first area of the hindbrain). These brain regions in mammals, most notably the cerebrum, are known to be responsible for many higher functions, including increased sensory functions.

Mobulid rays also have a network of blood vessels called the rete mirabile, or 'wonderful net', which surrounds their braincase and retains heat via a countercurrent exchange mechanism, helping to keep the brain warmer than the surrounding tissue. When manta rays dive to depths of over 750 metres (2,460 feet), or a Sicklefin Devil Ray *Mobula tarapacana* dives to a massive 2,000 metres (65,640 feet), where temperatures rapidly decrease, this rete system is likely to be important, allowing their brains to keep functioning effectively even when the external temperature of the surrounding water is much colder. Mammals and birds can also thermoregulate – indeed, it is one of the major reasons why these animal groups have been so successful.

Whether or not these similarities between the brain physiology of mobulid rays and mammals is an indication of similar function and intelligence is not known, but mobulids' high degree of social interactions and curiosity towards humans would certainly suggest there is a lot more going on behind those eyes than is currently presumed. Recent behavioural studies on captive manta rays also suggest that they may be self-aware. If true, this would add further support to the claims that they are one of the most intelligent marine species.

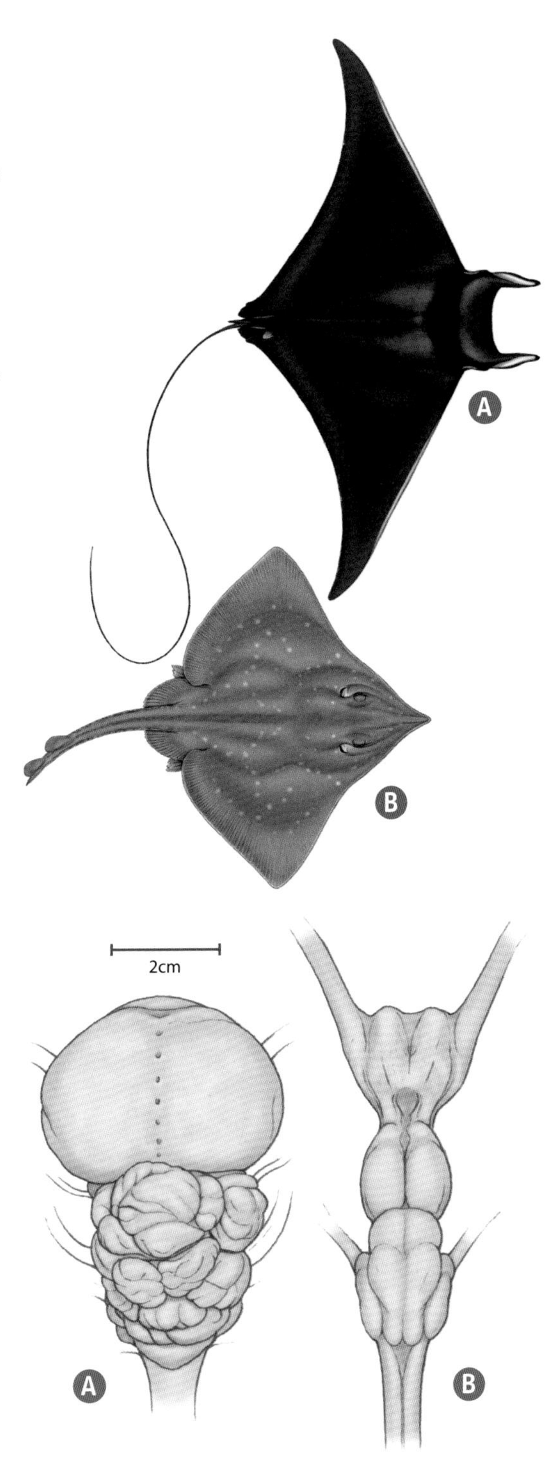

Left: Reef Manta Rays *Mobula alfredi* are social animals; they feed together in coordinated groups to maximise their feeding success.

Right: manta and devil rays have brains that are disproportionately large when compared to their body weight. The weight of the brain of a Spinetail Devil Ray *Mobula mobular* (a) for example, is more comparable to that of a similar-sized mammal – much larger than the brain of similar-sized relativse, like this Common Skate *Dipturus batis* (b).

Patterning and morphs

The coloration and spot patterns of manta and devil rays vary between species. However, to complicate matters, in manta rays they also vary within and between different populations of the same species. These variations range from one extreme to the other, so while it is not possible to clearly define manta rays through their body coloration and patterning alone, there are some generalities that most individuals conform to.

Within both species of manta there is one main colour form, or morph, which usually dominates the population. This is called the 'chevron morph' (see pages 54–57). Chevron mantas are characterised by a black dorsal (topside) surface that has two white, triangular-shaped patches positioned across the top of the head, while the ventral (underside) surface is white in colour with a varying degree of black spots and shaded patterning spread across the whole ventral area. Some of these chevron mantas have virtually no spots, while others are covered in hundreds. Furthermore, in some individuals the white dorsal shading can extend to cover the entire surface of the manta's back, giving it a ghostly white appearance. These individuals are known as leucistic colour morphs and are relatively common in some populations, such as the Maldives Reef Manta, but rare elsewhere.

The other morph seen in some manta populations around the world is the melanistic or black manta. In these individuals the back and most of the ventral surface are completely black, with usually a small patch of white left around the gill areas on the underside. These black mantas look as though a chevron manta has been dipped in black paint upside down, leaving just the smallest areas undipped. Indeed, if you look closely you can see the underlying darker black spots that would have been more visible if the individual's chevron coloration was not overridden by the black pigmentation.

Black or melanistic (top right) and chevron (right) morph Reef Manta Rays *Mobula alfredi*.

Like an underwater stealth bomber, a black morph Oceanic Manta Ray *Mobula birostris* slowly circles a seamount called Roca Partida in the Revillagigedo Archipelago, Mexico.

These melanistic mantas are unbelievably striking and the black Oceanic Mantas are the kings of all. Like marine stealth bombers they cruise through the oceans with a sense of grace that never ceases to captivate. At the remote islands of Socorro, San Benedicto and Roca Partida in the Revillagigedo Archipelago, 400 kilometres (250 miles) southwest of the southern tip of Baja California, a higher percentage of these giant oceanic blacks can be found than anywhere else in the world. The mantas visiting these isolated islands and seamounts are often extremely inquisitive and playful, ensuring some breathtaking and personal encounters between humans and these animals.

Colour and spot differences between the Reef and Oceanic species are subtle. In general, there are more acute transitions between the black and white areas on the dorsal surface of the Oceanic Mantas, and an absence of spots from the whole of the ventral surface except for a small central cluster usually present near the tail on the Oceanic Manta Ray's belly. These subtle differences make it hard for the untrained eye to differentiate between the Reef Mantas and the Oceanic Mantas based on visual observations alone – especially when variations within and between populations are often much more pronounced than colour variations between the two species.

Although confusing from a layman's perspective, these variations in spot patterning are extremely useful for manta scientists seeking to identify each individual manta they study. Each manta's spots are unique, just like a human fingerprint. It is born with these spots and they remain unchanged throughout its life. Manta scientists can therefore build photo-ID databases of all the mantas in their population. This allows them to estimate the population number, their movements (seasonally and spatially) and the overall demographics and reproductive cycles of the population.

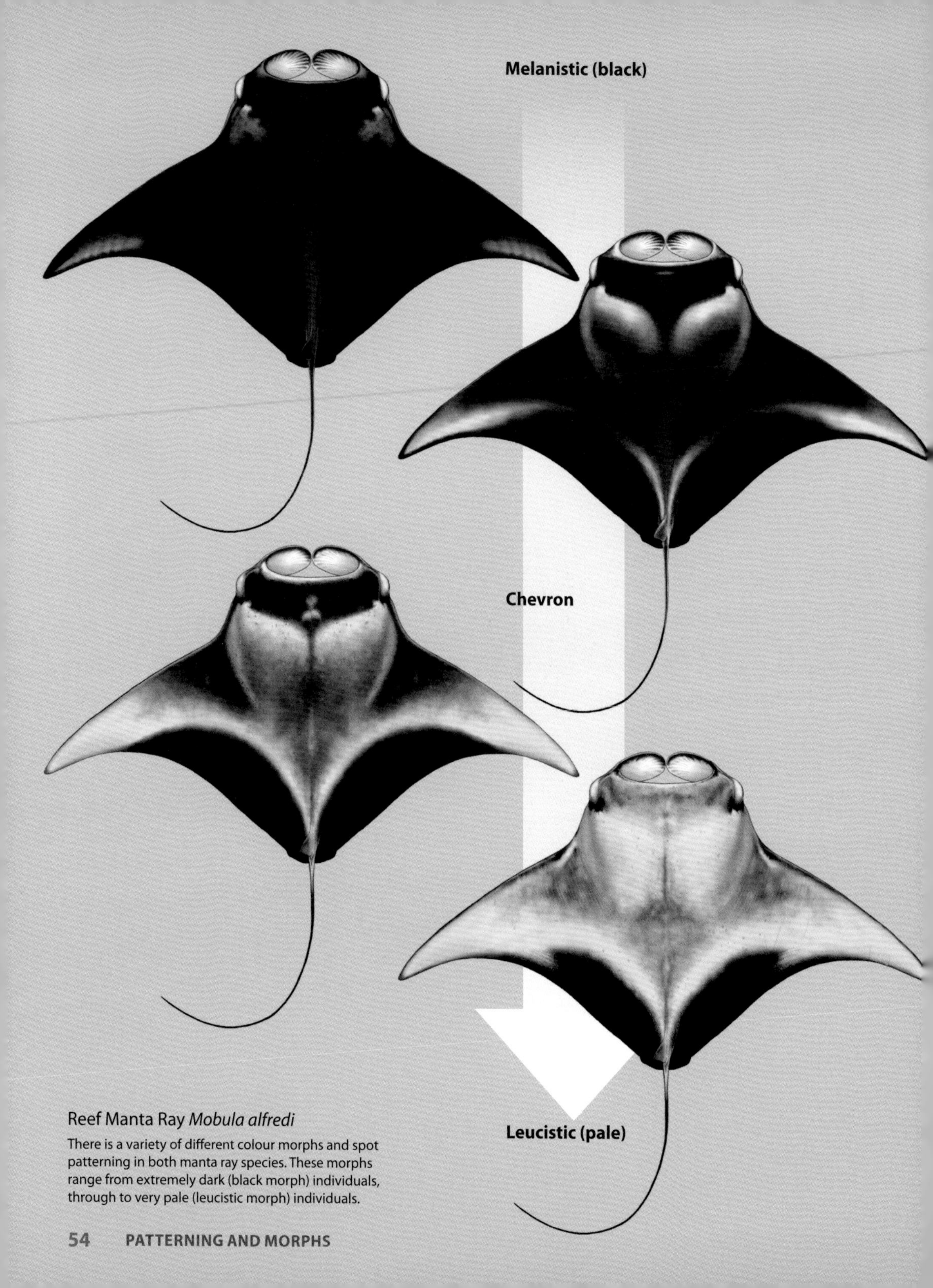

Reef Manta Ray *Mobula alfredi*

There is a variety of different colour morphs and spot patterning in both manta ray species. These morphs range from extremely dark (black morph) individuals, through to very pale (leucistic morph) individuals.

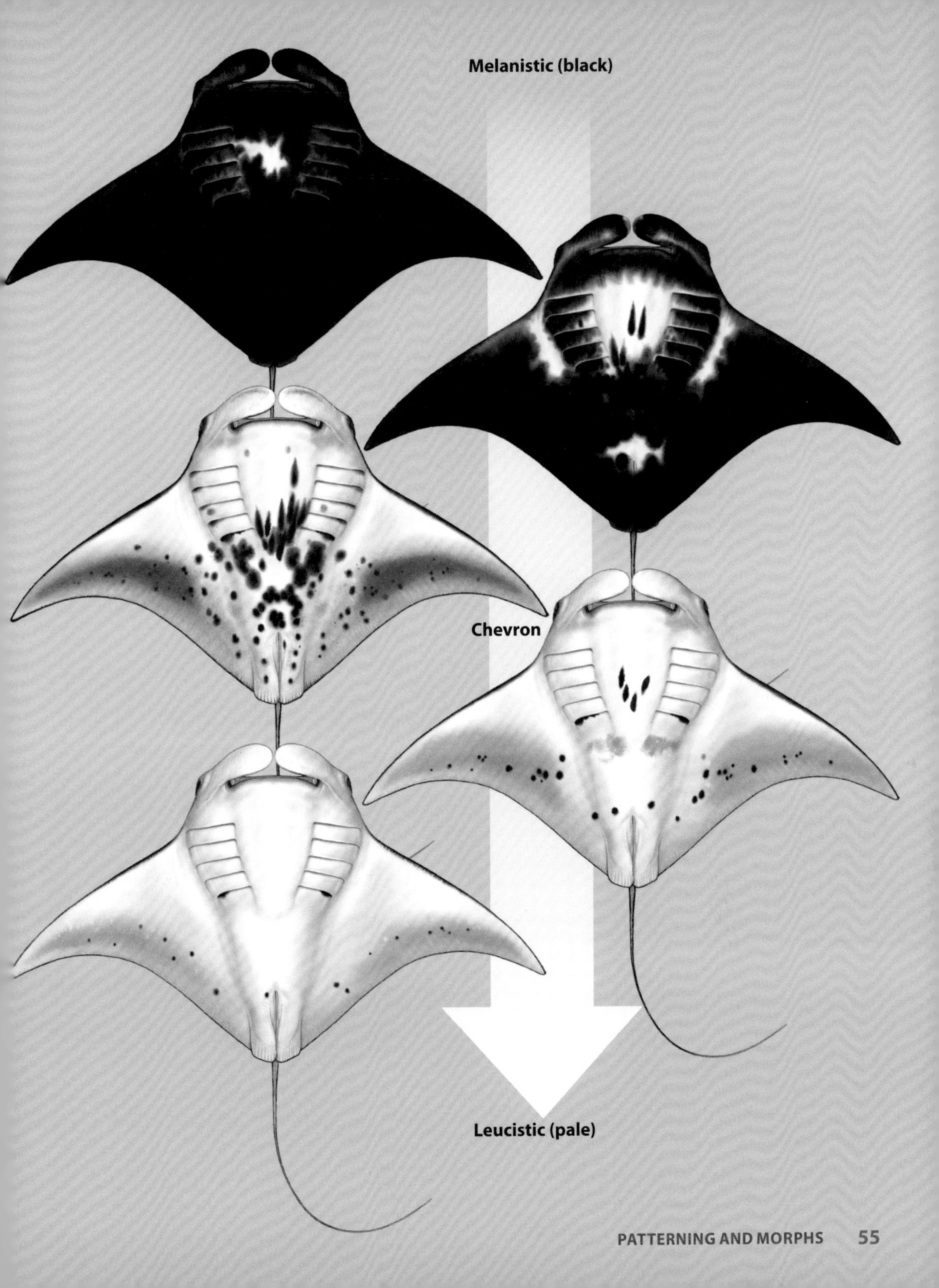
Melanistic (black)
Chevron
Leucistic (pale)

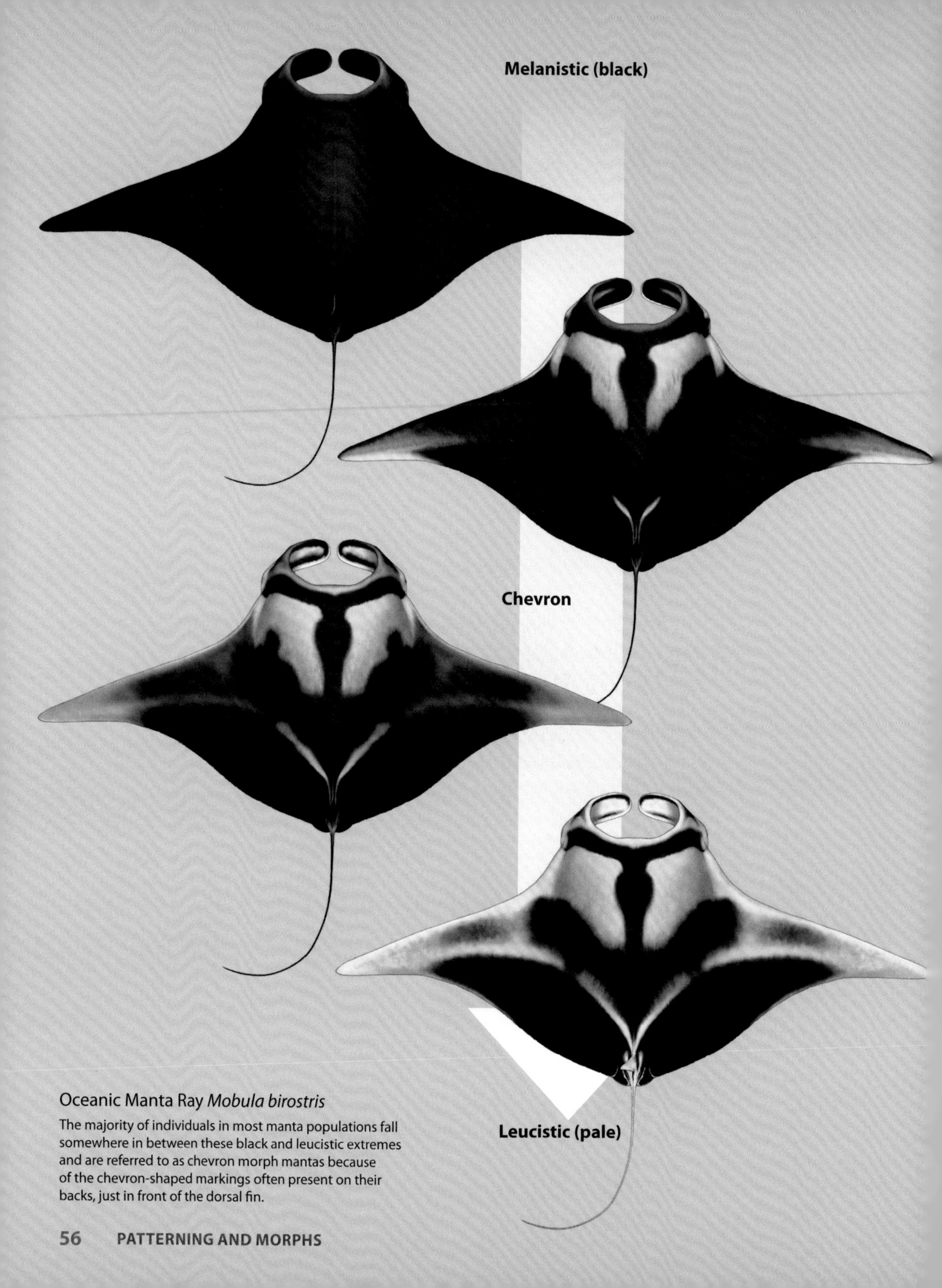

Oceanic Manta Ray *Mobula birostris*

The majority of individuals in most manta populations fall somewhere in between these black and leucistic extremes and are referred to as chevron morph mantas because of the chevron-shaped markings often present on their backs, just in front of the dorsal fin.

Melanistic (black)
Chevron
Leucistic (pale)

Morphology

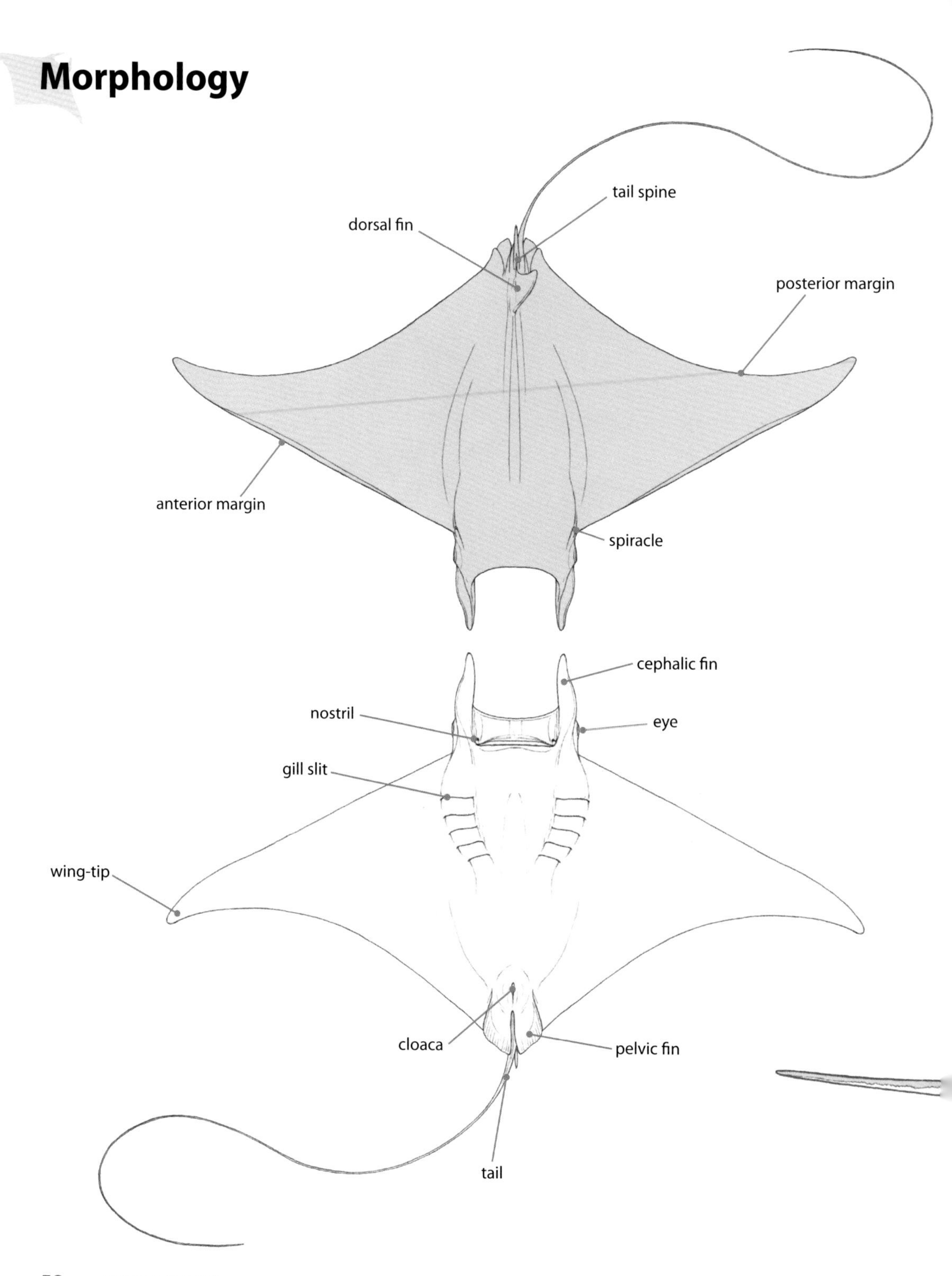
tail spine
dorsal fin
posterior margin
anterior margin
spiracle
cephalic fin
nostril
eye
gill slit
wing-tip
cloaca
pelvic fin
tail

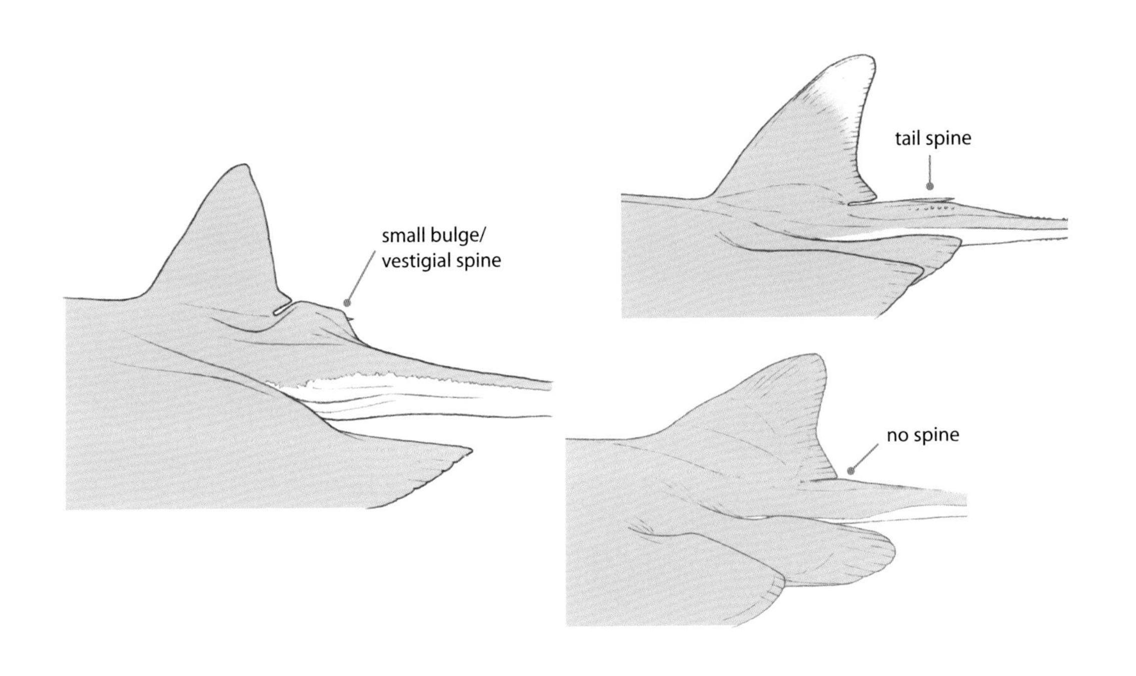
small bulge/
vestigial spine
tail spine
no spine

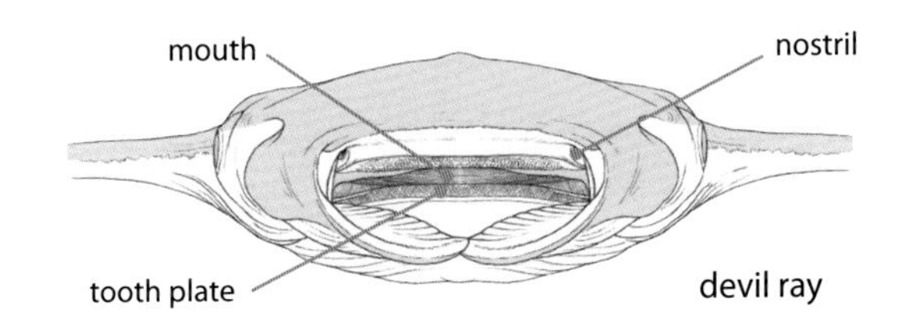
mouth
nostril
tooth plate
devil ray

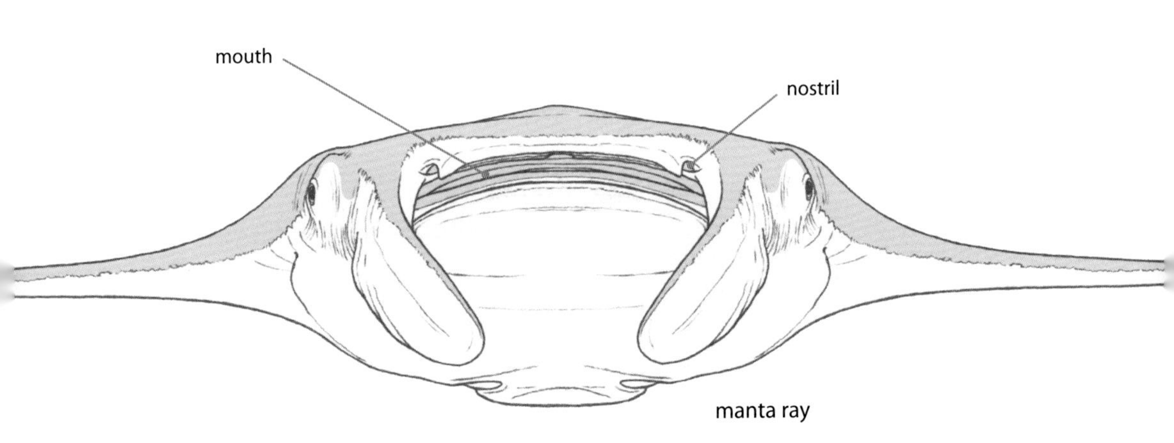
mouth
nostril
manta ray

Spiracles

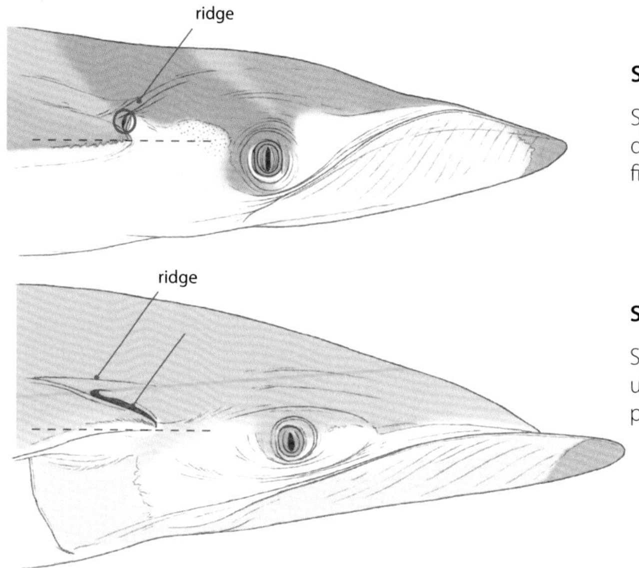

Spinetail Devil Ray *Mobula mobular*

Spiracle is a short transversal slit present under a distinct ridge, above the margin of the pectoral fin near where the fin meets the body.

Sicklefin Devil Ray *Mobula tarapacana*

Spiracle in an elongated longitudinal slit under a ridge above and behind the margin of pectoral fin where it meets the body.

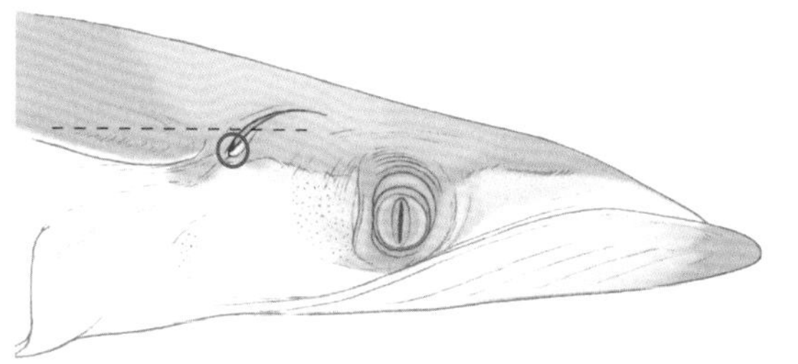

Shorthorned Pygmy Devil Ray *Mobula kuhlii*

Spiracle very small, subcircular and below the margin of the pectoral fin where it meets the body.

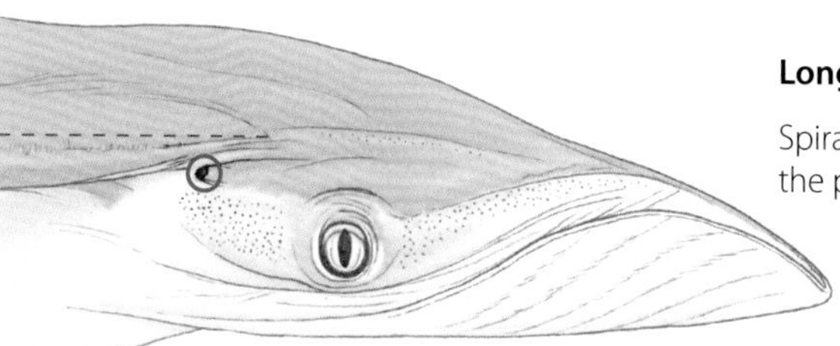

Longhorned Pygmy Devil Ray *Mobula eregoodootenkee*

Spiracle very small, subcircular and below the margin of the pectoral fin where it meets the body.

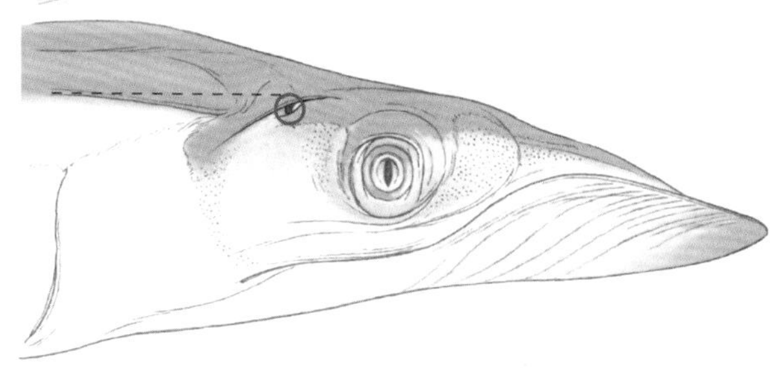

Bentfin Devil Ray *Mobula thurstoni*

Spiracle small, subcircular and below the margin of the pectoral fin where it meets the body.

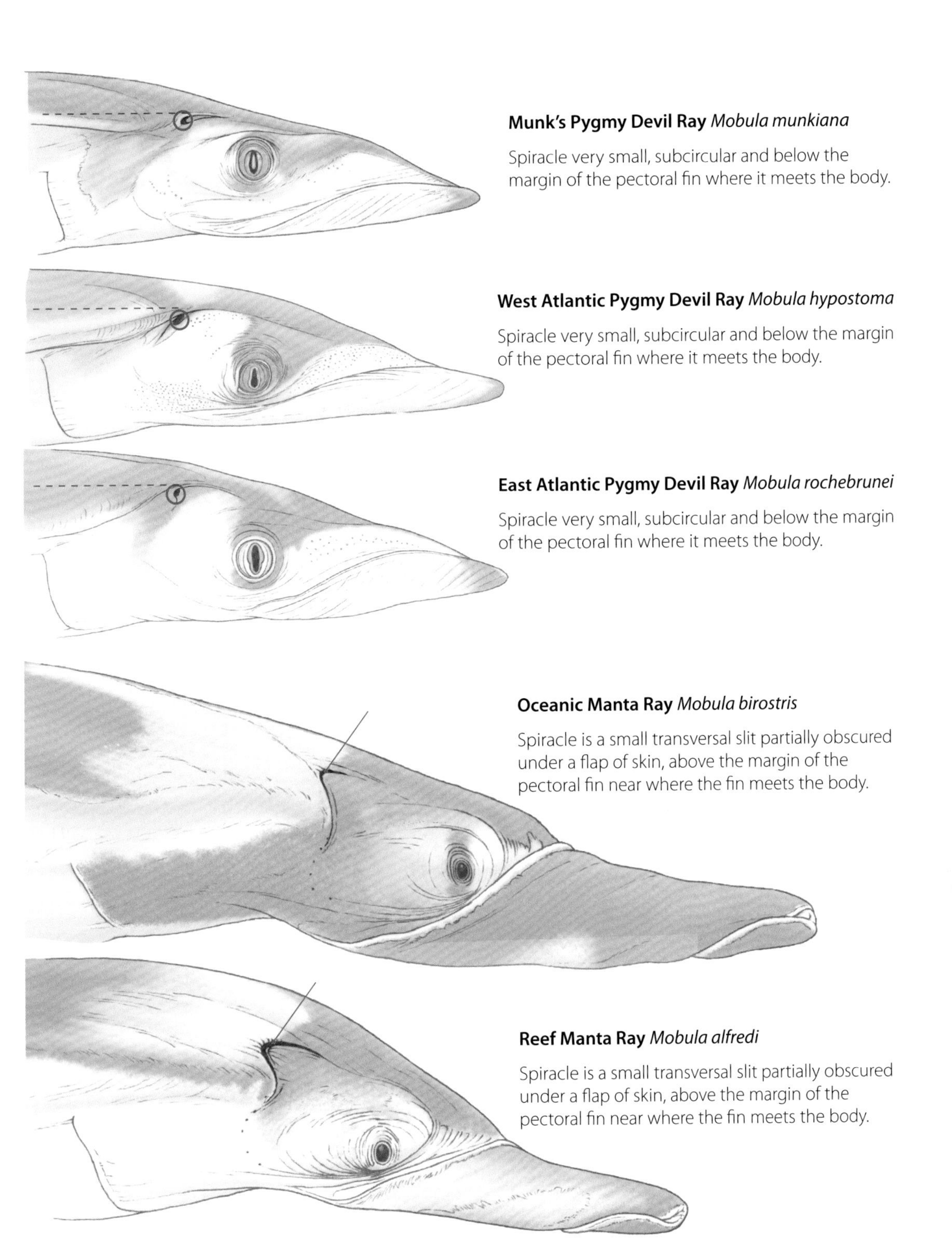
Munk's Pygmy Devil Ray *Mobula munkiana*
Spiracle very small, subcircular and below the margin of the pectoral fin where it meets the body.
West Atlantic Pygmy Devil Ray *Mobula hypostoma*
Spiracle very small, subcircular and below the margin of the pectoral fin where it meets the body.
East Atlantic Pygmy Devil Ray *Mobula rochebrunei*
Spiracle very small, subcircular and below the margin of the pectoral fin where it meets the body.
Oceanic Manta Ray *Mobula birostris*
Spiracle is a small transversal slit partially obscured under a flap of skin, above the margin of the pectoral fin near where the fin meets the body.
Reef Manta Ray *Mobula alfredi*
Spiracle is a small transversal slit partially obscured under a flap of skin, above the margin of the pectoral fin near where the fin meets the body.

Gill plates

Mobulid pre-branchial gill plates should NOT be called gill rakers. Gill plates are cartilage-supported appendages, and as such are part of the endoskeleton. By contrast, gill rakers are exoskeletal dermal plates found in some bony fishes (e.g. anchovies and herring) that may or may not carry teeth, and they ossify in the dermis without any cartilage precursors.

Each mobulid ray has five pairs of gill arches (the skeletal structures supporting the gills and gill plates), each of which is encircled internally by a prebranchial appendage, or gill plate. These structures act as filters, straining the planktonic food from the water column, enabling manta and mobulid rays to feed.

All mobulid species have slightly different gill plate structures; each adapted to improve the capture of their target prey. For example, those species which feed on small fishes, such as the Sicklefin Devil Ray *Mobula tarapacana* and Longhorned Pygmy Devil Ray *Mobula eregoodootenkee*, have coarse plates to allow smaller particles to pass through, deflecting the larger fish prey into the back of the devil ray's mouth for ingestion. The manta rays, however, have much finer feathery gill plates, deflecting their smaller zooplankton prey backwards into the buccal cavity of the animal.

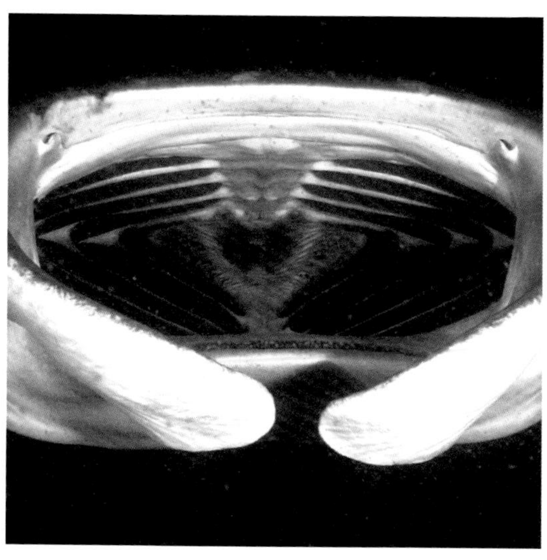

The cavernous mouth of a manta ray is encircled by five pairs of gill plates, which funnel their zooplankton prey towards their throat.

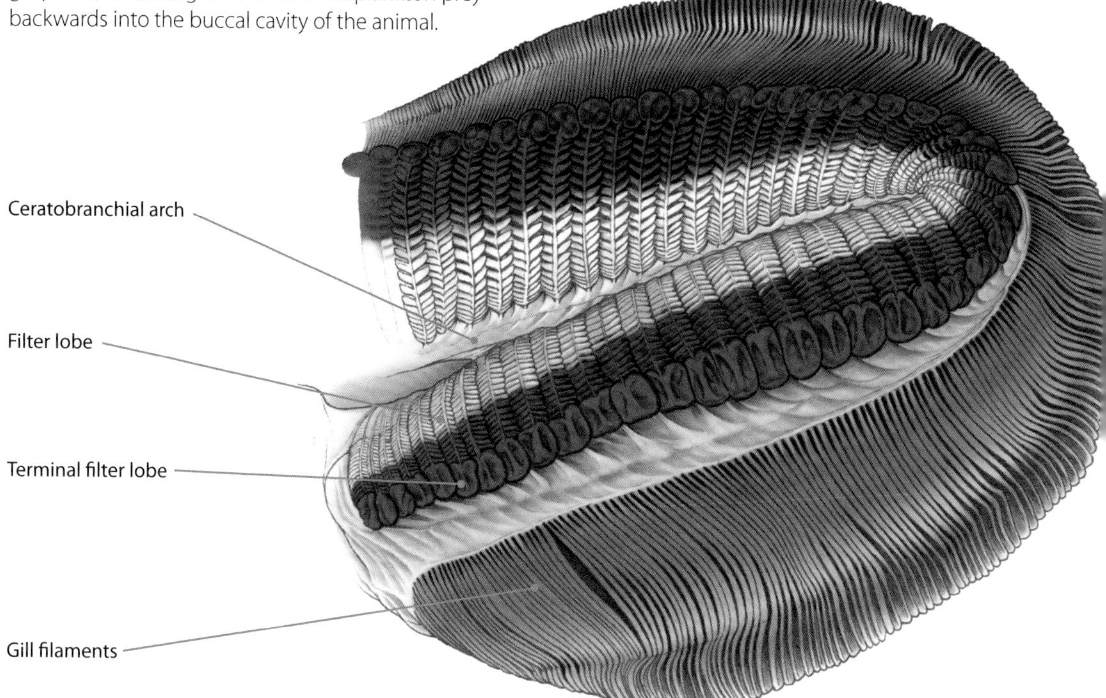

Mobula alfredi

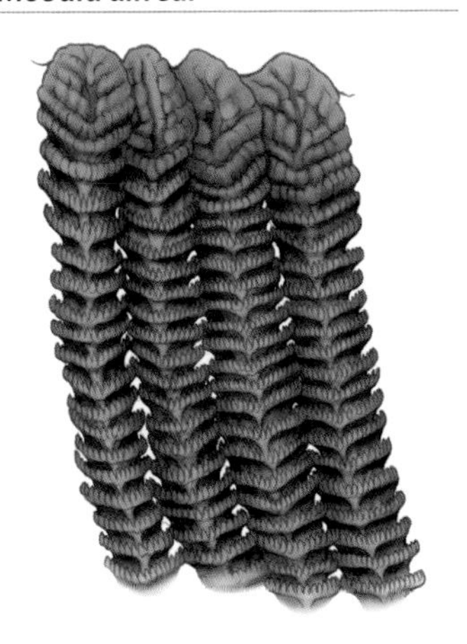

Mobula birostris

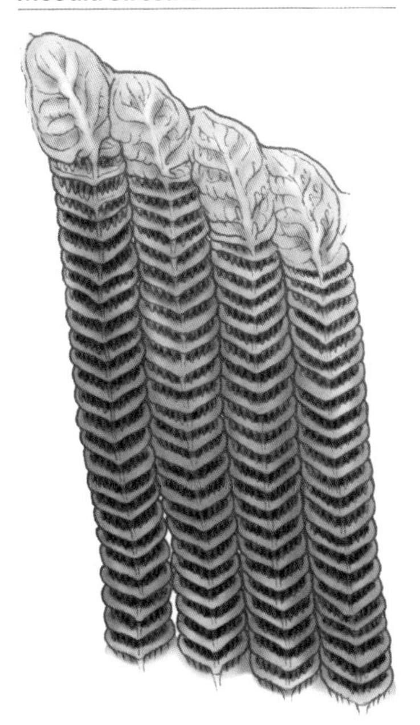

Mobula mobular

Mobula eregoodootenkee

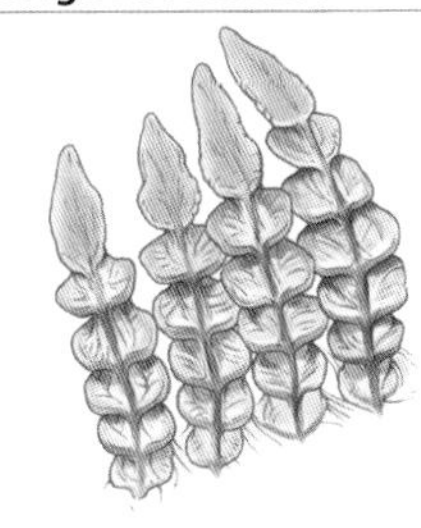

Mobula munkiana

Mobula hypostoma

Mobula kuhlii

Mobula rochebrunei

Mobula tarapacana

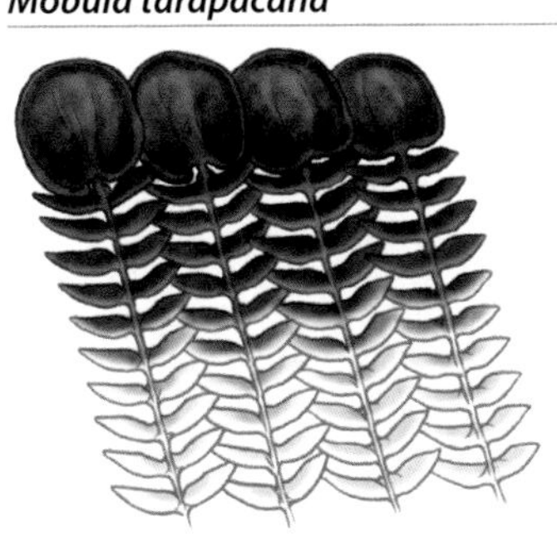

Mobula thurstoni

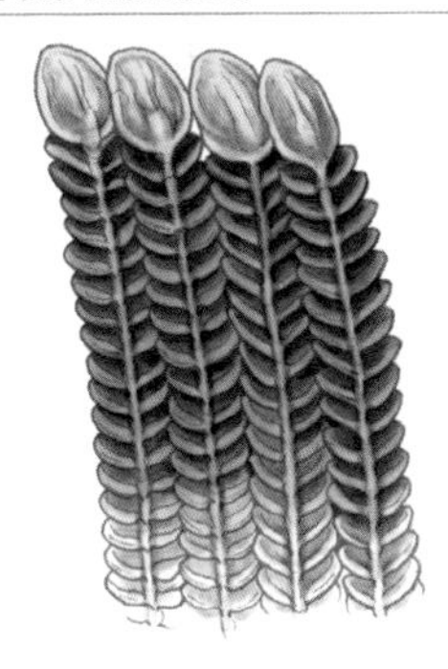

Sexual dimorphism

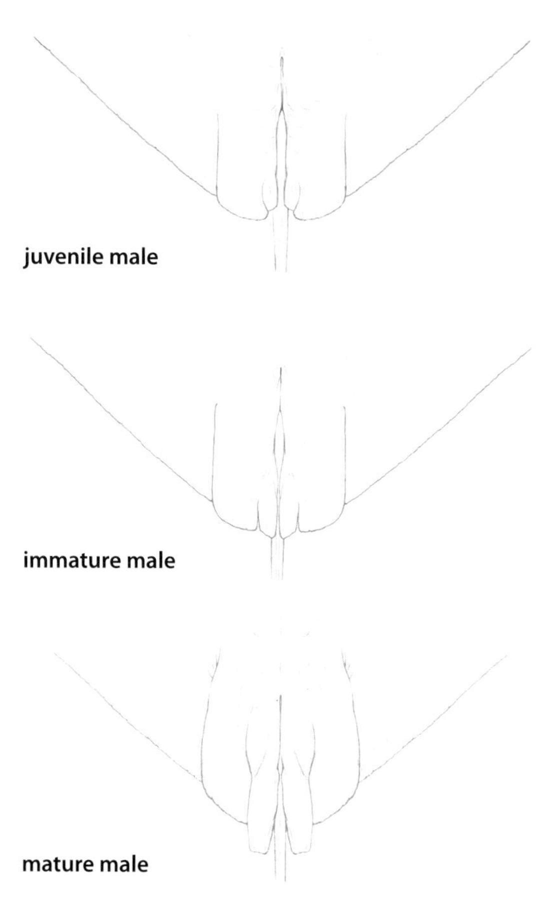

In every mobulid species females always attain larger disc widths than the males (above – male on top). This sexual dimorphism is most likely an adaptation that enables the females to carry and give birth to large pups, which are much more likely to survive in the wild on their own than smaller offspring.

The maturity of female manta and devil rays can be determined in the wild through the presence of mating scars and visible pregnancies. Maturity of the male mobulid rays, like all sharks and rays, can be noted instead by the size and appearance of their sexual organs, the claspers. When a male reaches sexual maturity his small, soft and pliable claspers become enlarged, calcified and hardened for use in copulation (below). This allows scientists studying these animals in the wild to record and document the males' maturity as they develop from juveniles, to subadults, to fully mature males.

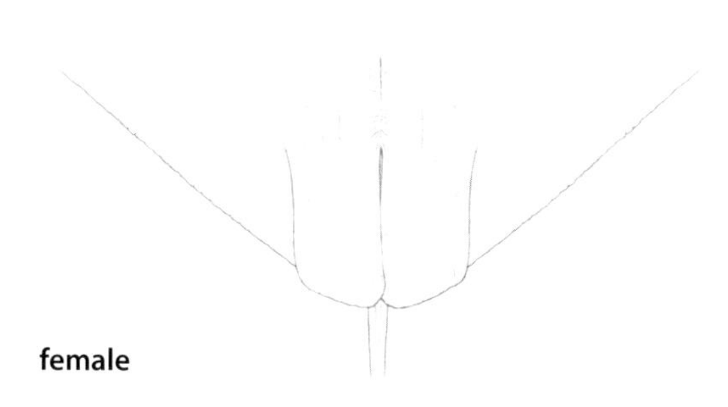

SPECIES ACCOUNTS

Species key

Munk's Pygmy Devil Rays *Mobula munkiana*, Cabo San Lucas, Mexico.

1a

Terminal mouth; head width 21–22% of DW (disc width); toothband present only on lower jaw.

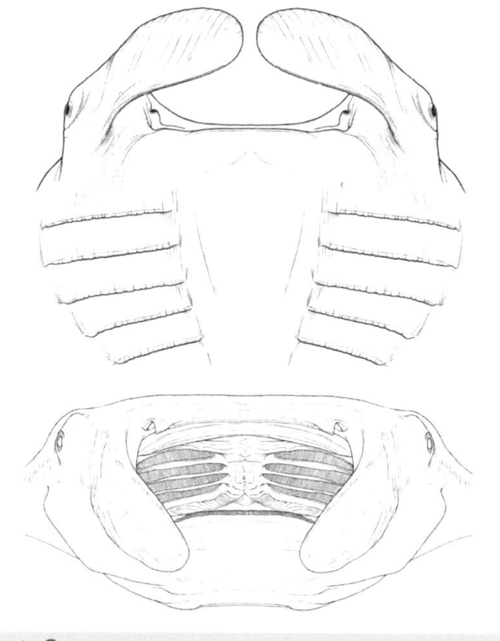

→ **2**

1b

Ventral (undercut) mouth; head width 16–17% of DW (disc width); toothbands in both jaws.

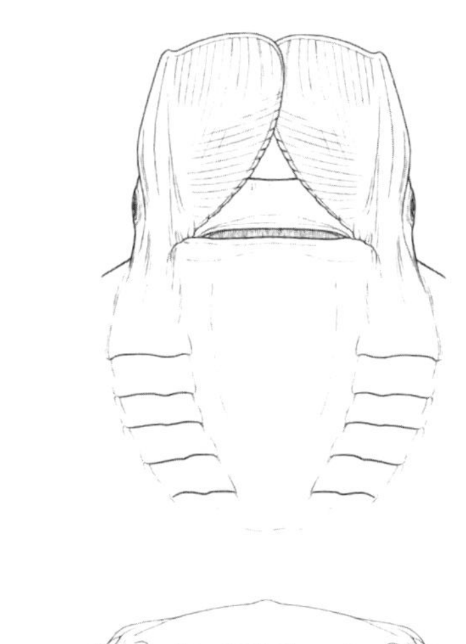

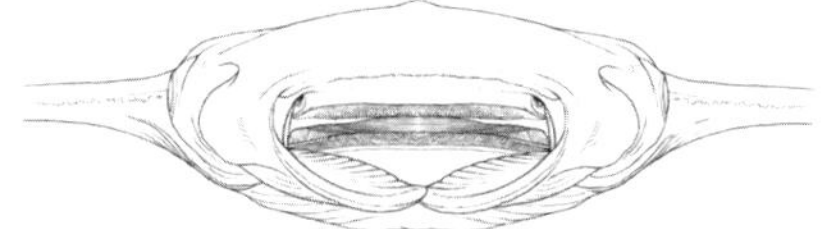

→ **3**

2

If present, ventral spots clustered around lower abdominal region only (1); gill covers (particularly 5th gill) and mouth with black shading/flaring (2); dorsal white shoulder markings form two mirror image right-angled triangles creating a 'T' in black*. Large: adult size DW>400cm.

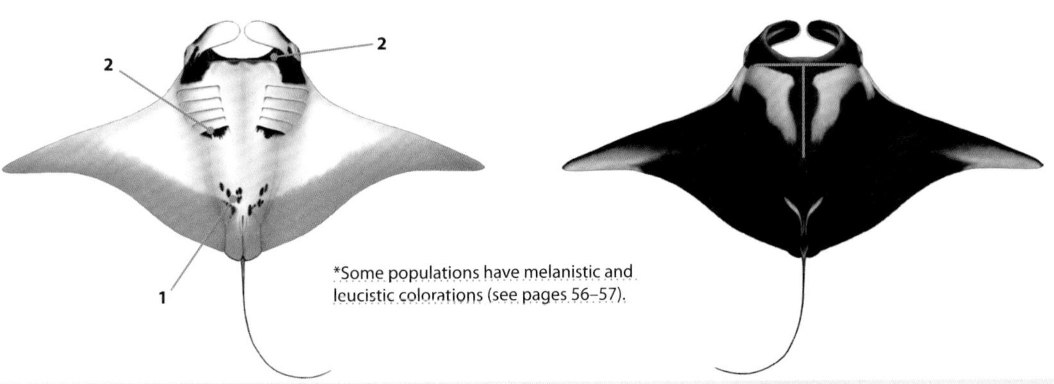

YES → ***Mobula birostris*** (found circumtropically throughout the Indo-Pacific and Atlantic Oceans) **page 74**

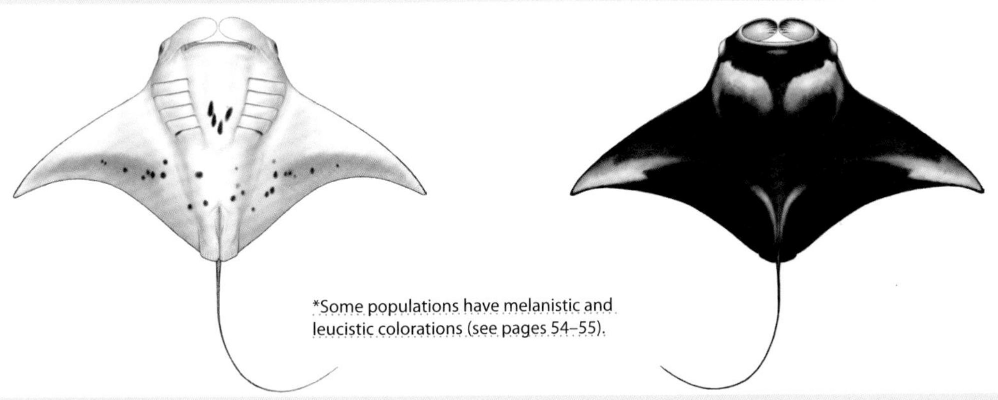

NO → ***Mobula alfredi*** (found throughout the tropics of the Indo-West Pacific Oceans) **page 70**

3

White ventral markings wrap up behind and above the uppermost level of the eyes (1), and these white markings on either side are clearly visible when viewing the dorsal surface of specimen directly from above (2).

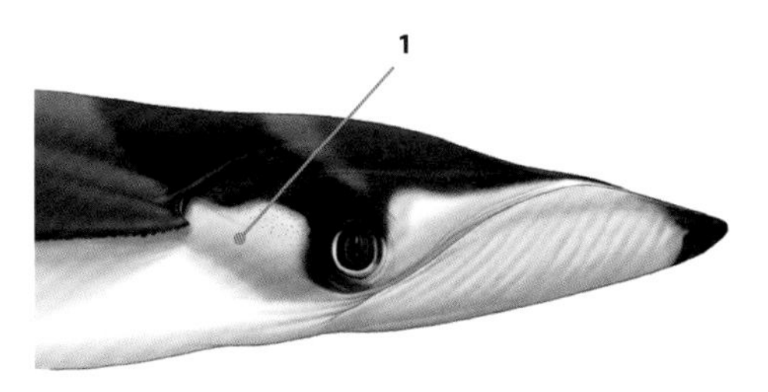

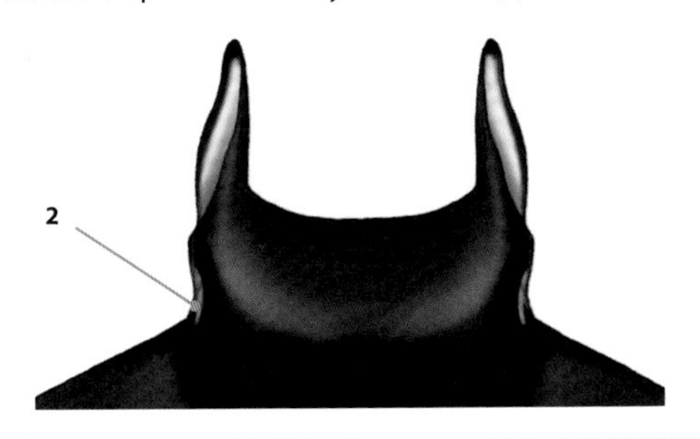

YES → **4**

NO → **7**

4

Caudal spine present; spiracle under a distinct ridge **above** the margin of the pectoral fin where it joins body (2); tail equal to or longer than disc width (3).

Large: adult size DW 320cm. Found circumtropically throughout the Indo-Pacific and Atlantic Oceans.

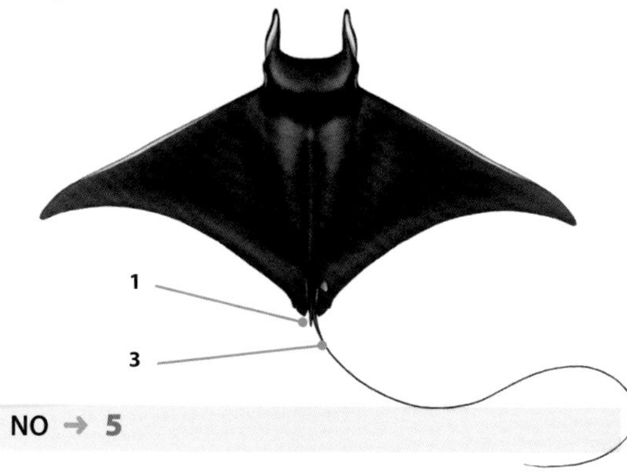

YES → *Mobula mobular* **page 90**

* *japanica* has merged with *mobular*

NO → 5

5

Found only in the Eastern Pacific Ocean. Small: adult size DW<150cm.

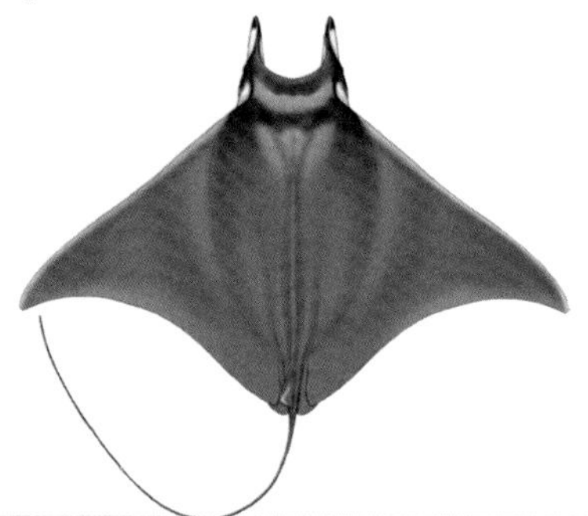

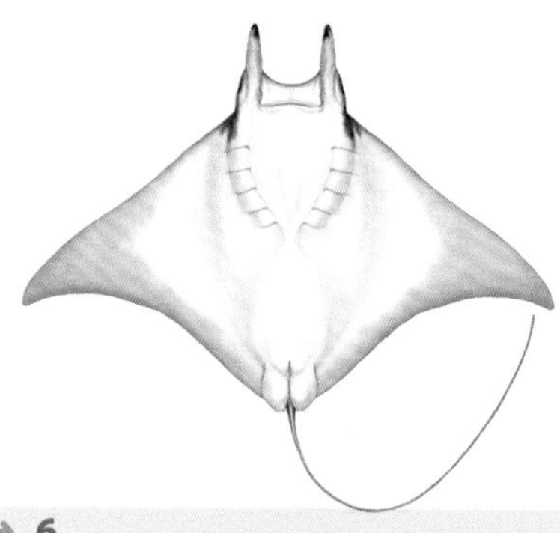

YES → *Mobula munkiana* **page 94**

NO → 6

6

Found only in the Western Atlantic Ocean.
Small: adult size DW<150cm.

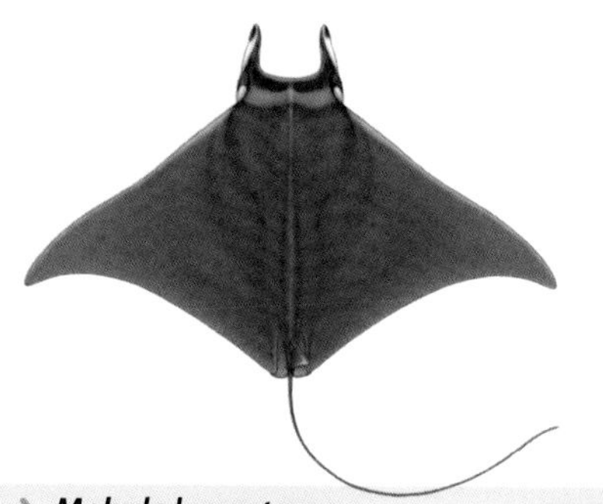

YES → *Mobula hypostoma* **page 82**

Found only in the Eastern Atlantic Ocean.
Small: adult size DW<150cm.

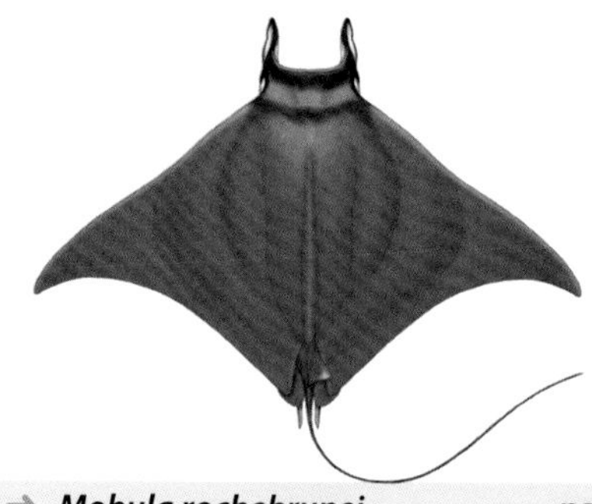

NO → *Mobula rochebrunei* **page 98**

7

Long-necked appearance; small head to body ratio (1); trailing edge of pectoral fins distinctly falcate (2); spiracle under a ridge **above** and behind the margin of pectoral fin where it joins body; dark grey shading on first (or more) gill cover(s) (3); grey ventral shading on posterior margin of pectoral fins and white anteriorly, with irregular demarcation between both (4); olive-green to brown dorsally. Large: adult size DW 340cm. Found circumtropically throughout the Indo-Pacific and Atlantic Oceans.

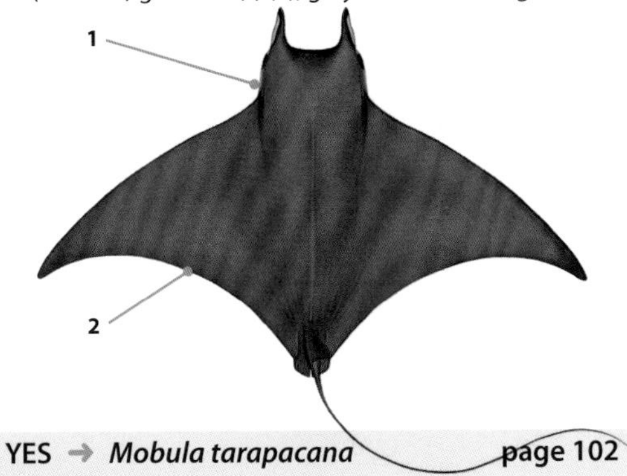

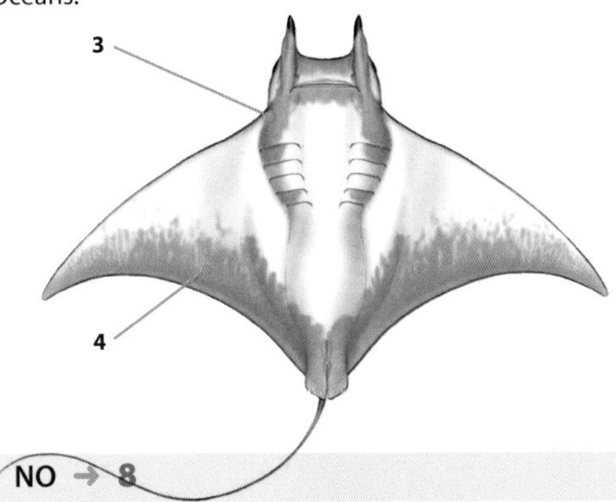

YES → *Mobula tarapacana* **page 102**

NO → 8

8

Medium-sized species reaching 183cm DW; anterior margin of pectoral fins have a distinctive double curvature (1) with golden black-grey shading on curve ventrally (2); large pelvic fins which extend past the base of the pectoral fins (3). Found circumtropically throughout the Indo-Pacific and Atlantic Oceans.

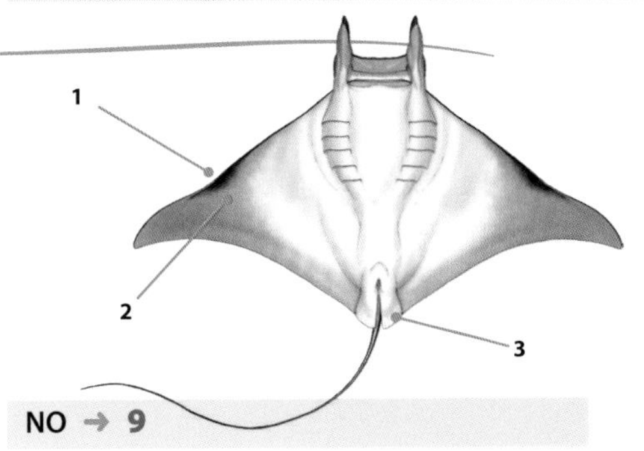

YES → *Mobula thurstoni* **page 106**

NO → 9

9

Long-necked appearance (1); distinct triangular-shaped black to dark-grey shading on the leading edge of pectoral fin at the mid-point (2); long cephalic fins with length from tip of each fin to corner of mouth greater than 16% DW (3). Small: adult size DW<150cm. Found throughout the Indo-West Pacific Oceans.

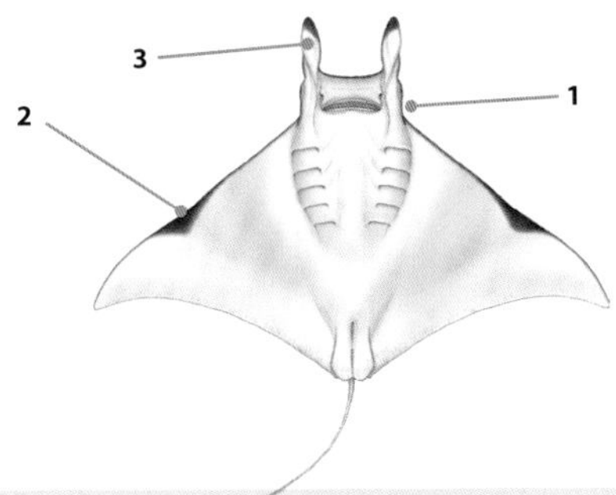

YES → *Mobula eregoodootenkee* **page 78**

Short-necked appearance (1); light grey stripe runs along the anterior dorsal margin of the pectoral fins (2); dark head collar (3); short cephalic fins with length from tip of each fin to corner of mouth less than 16% DW (4). Small: adult size DW<150cm. Found throughout the Indo-West Pacific Oceans.

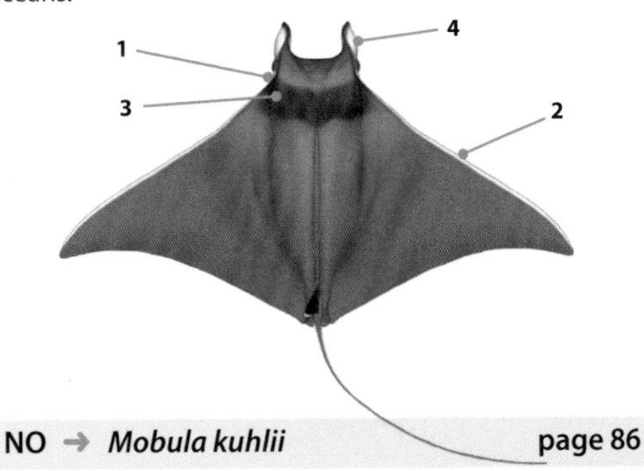

NO → *Mobula kuhlii* **page 86**

Reef Manta Ray *Mobula alfredi*

(Krefft, 1868)

Reef Manta Ray *Mobula alfredi*, Baa Atoll, Maldives.

Species characteristics

Disc width: maximum 450cm (15ft), average 300–350cm (10–11.5ft).

Weight: up to 700kg (1,500lbs).

Size at maturity: ♀ 320–350cm (10.5–11.5ft), ♂ 270–300cm (9–10 ft).

Age at maturity: ♀ ~15 years, ♂ ~9 years.

Lifespan: likely to be around 40 years.

Reproduction: one live-born pup (rarely two) on average every 2–5 years.

Distribution: Indo-West Pacific, to 32°N (Shikoku, Japan and Red Sea, Egypt) and 30°S (Margate, South Africa and New South Wales, Australia).

IUCN Red List: **Vulnerable**

FAO species code: **RMA**

The Reef Manta Ray is one of the largest and most iconic marine species. It has an average disc width (wing tip to tip) for adults of about 300–350cm (10–11.5ft), depending on the location and sex of the animal. Because Reef Mantas frequent the relatively shallow waters along the coastal reefs of continents and oceanic islands they are more commonly encountered by divers and snorkellers than their Oceanic Manta Ray cousins. These smaller mantas are highly social and often resident to a specific home range, migrating around these areas as they follow changes in the seasonal abundance of their planktonic food source. They frequent the same sites year after year for many decades, allowing researchers to gather in-depth data on the population as a whole and to follow more closely the lives of specific individuals as they grow, reproduce and migrate.

Reef Mantas spend much of their time visiting cleaning stations on coral reefs, allowing small fishes to scour every surface of their bodies for morsels of food. These cleaner fish are looking for parasites lodged in the mantas' mouths, gills and on their skin. These cleaning stations play a key role in the lives of the mantas, acting as important social gathering points

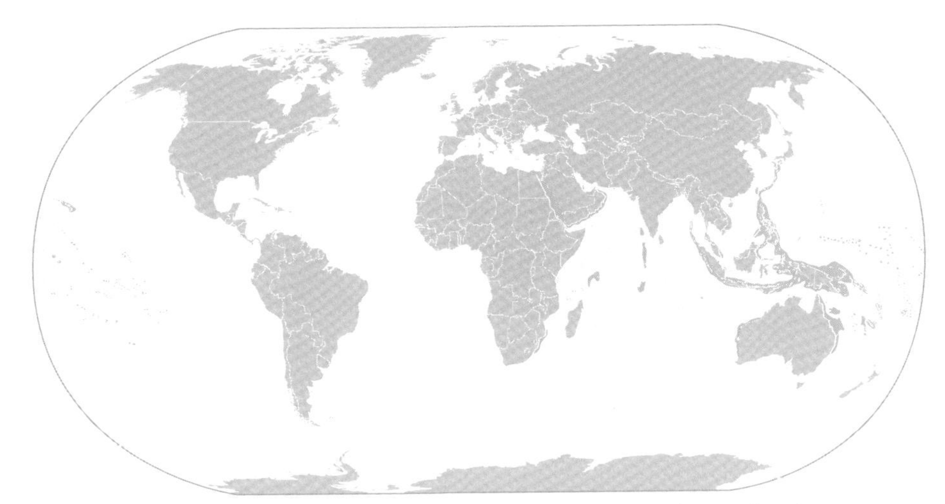

Distribution map of the Reef Manta Ray *Mobula alfredi*. Darker areas indicate confirmed range; lighter areas indicate expected range.

for courting and mating. They also serve as a refuge from predators and possibly as a warm place to thermoregulate, aiding digestion and gestation.

Both manta species are threatened throughout their range by targeted fisheries driven by the demand for their gill plates, which are used in Asian medicines. Bycatch fisheries are also a threat to the Reef Manta Ray, which easily becomes entangled in nets and fishing line. Climate change is likely to effect the abundance of the zooplankton prey upon which this species relies.

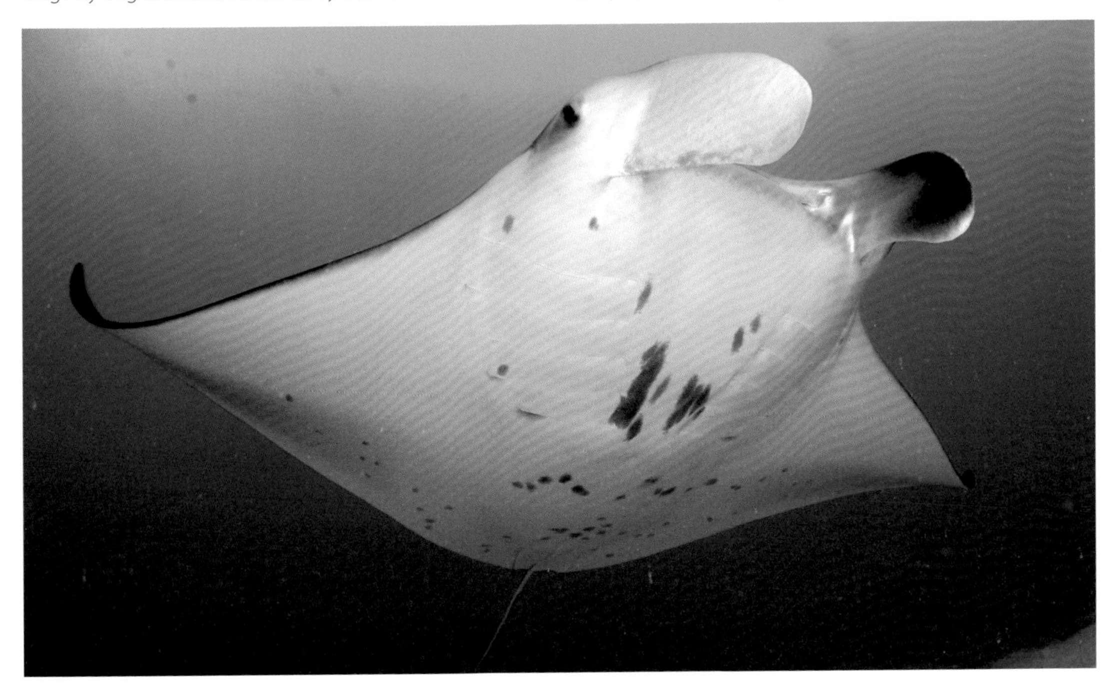

A Reef Manta Ray *Mobula alfredi* in the Amirantes, Seychelles, shows its belly to the photographer. Each manta's spots are unique, just like a human fingerprint, they remain unchanged throughout its life. Manta scientists can therefore build databases of all the mantas in their population, following their lives over the decades.

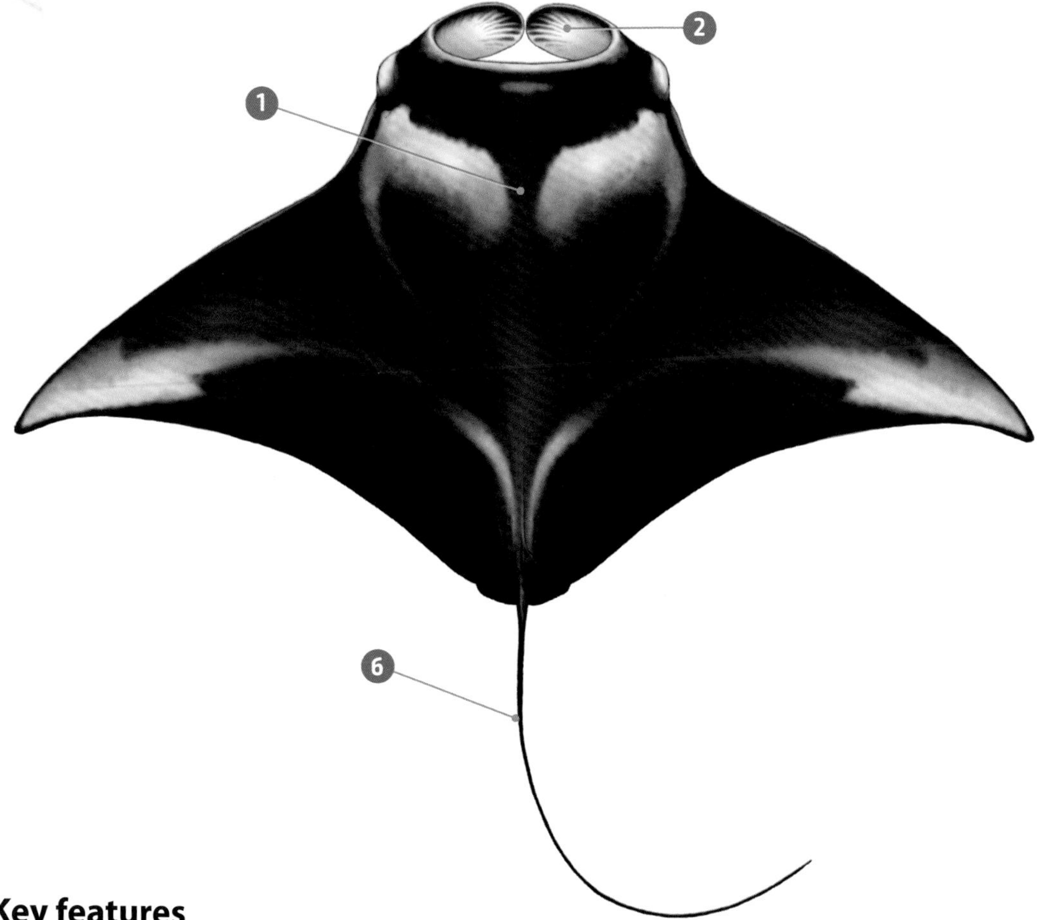

Key features

1. Transition between white and black markings on dorsal surface blurred along colour boundary forming more of a 'V' in black.
2. Inside of mouth and cephalic fins, and the trailing underside edge of pectoral fins usually shaded grey or white (excluding melanistic morph).
3. If present, ventral spots often between branchial gill slits and across trailing edge of pectoral fins and abdominal region.
4. Gill covers (particularly 5th gill) occasionally with grey shading/flaring.
5. Uniform dorsal fin with no white tip.
6. Tail equal to or shorter than its disc width.
7. Slight depression at base of tail, although occasionally a small bulge and vestigial spine is present.
8. Cephalic fins large, unfurling to meet together in centre of mouth.
9. Spiracle is a small, transversal slit partially obscured under a flap of skin, above margin of pectoral fin near where it meets the body.
10. Large gill plates with fused lateral lobes and rounded terminal lobe. Plates are coloured uniformly black, although occasionally completely white.

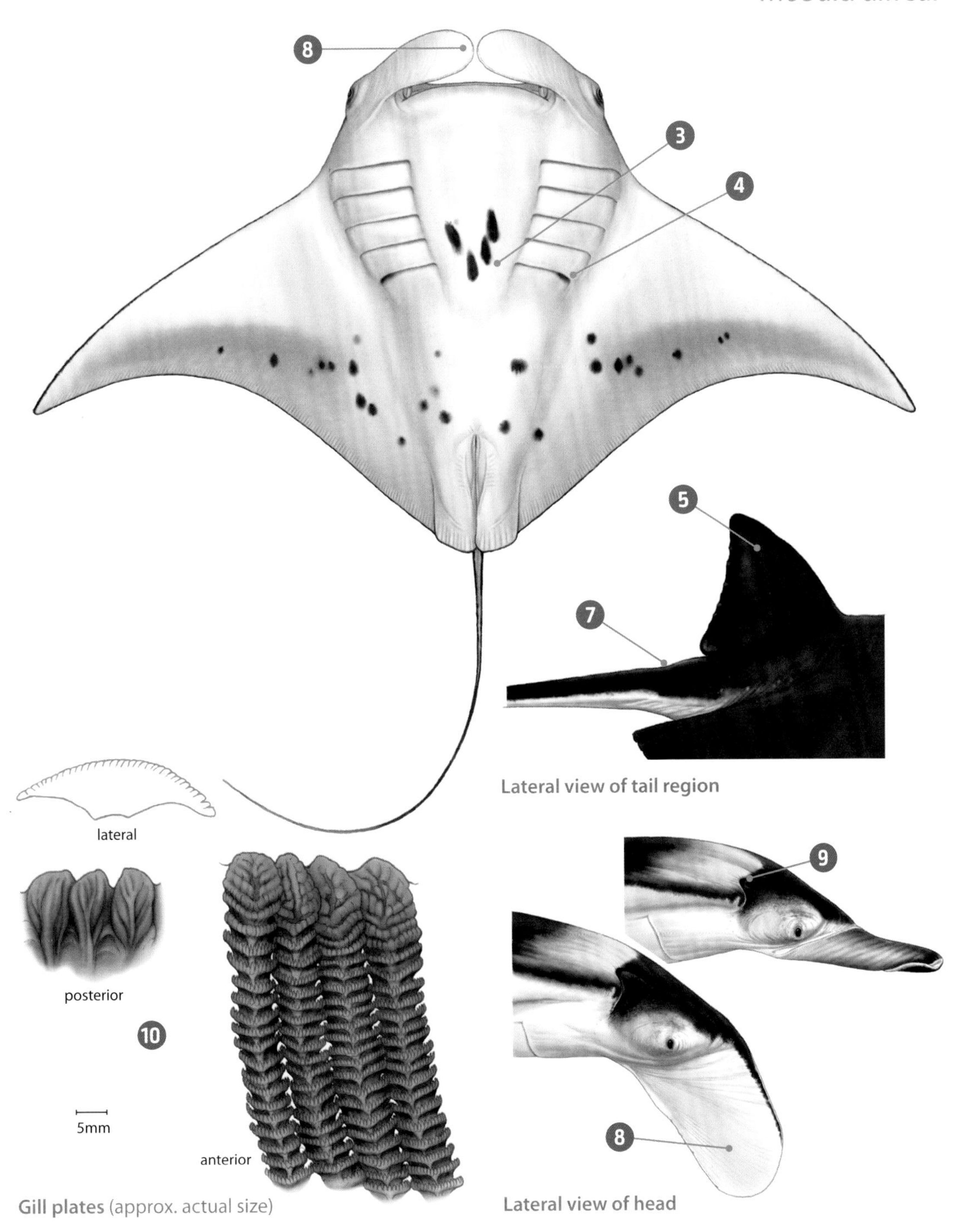

Gill plates (approx. actual size)

Oceanic Manta Ray *Mobula birostris*

(Walbaum, 1792)

Oceanic Manta Ray, *Mobula birostris*, Revillagigedo Archipelago, Mexico.

Species characteristics

Disc width: maximum 700cm (23ft), average 400–500cm (13–16.5ft).

Weight: up to 2,000kg (4,400lbs).

Size at maturity: ♀ 450–500cm (15–16.5ft), ♂ 350–400cm (11.5–13ft).

Age at maturity: unknown, but likely to be similar to Reef Manta Rays.

Lifespan: likely to be around 40 years.

Reproduction: unknown, but likely to be similar to Reef Manta Rays.

Distribution: found circumtropically in all oceans, to 40°N (New Jersey, USA and Honshu, Japan) and 40°S (northern Tasmania, Australia).

IUCN Red List: **Vulnerable**

FAO species code: **RMB**

The Oceanic Manta Ray is the largest ray species in the world, growing to a much larger size than its reef relative, with a disc width up to 700cm (23ft) in extremely large specimens, although the average size for most individuals encountered is usually around 400–500cm (13–16.5ft). There are also very slight morphological differences between the two species, including their teeth, skin texture and body patterning.

Oceanic Manta Rays appear to spend much of their time in the open ocean away from reefs, diving hundreds of metres into the deep scattering layer to find their zooplankton prey. They also spend a lot less time visiting cleaning stations than Reef Mantas. This elusive nature means we know a lot less about these giants than their smaller reef-dwelling relatives. We think these pelagic voyagers may follow the highway of ocean currents, moving in search of seasonal planktonic blooms. Until recently it was thought that Oceanic Manta Rays were ocean wanderers, travelling thousands of kilometres in search of food, possibly crossing entire ocean basins. However, recent studies suggest they are much more localised in their movements.

In 2009 manta rays were officially split into two distinct species based on new genetic and morphology studies, resulting

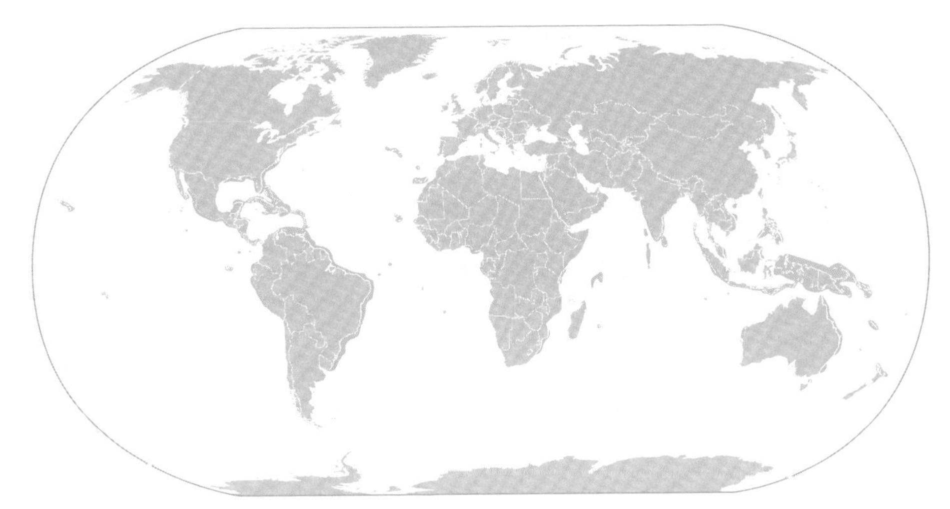

Distribution map of the Oceanic Manta Ray *Mobula birostris*. Darker areas indicate confirmed range; lighter areas indicate expected range.

in overlapping ranges of the two species throughout much of their range. In the Americas only the Oceanic Manta has been recorded. However, some scientists have proposed that a third species of manta ray may exist in the Caribbean (page 110). While there are some habitual and coloration differences between this Caribbean Manta Ray and the Oceanic Manta, there remains some uncertainty on the validity of this proposal based on genetic analysis.

An Oceanic Manta Ray *Mobula birostris* at the dive known as 'The Boiler' off San Benedicto Island in the Revillagigedo Archipelago, Mexico. This site is famous for Oceanic Manta Rays, which visit this site to get cleaned and to socialise with other manta rays. These inquisitive rays also like to check out divers.

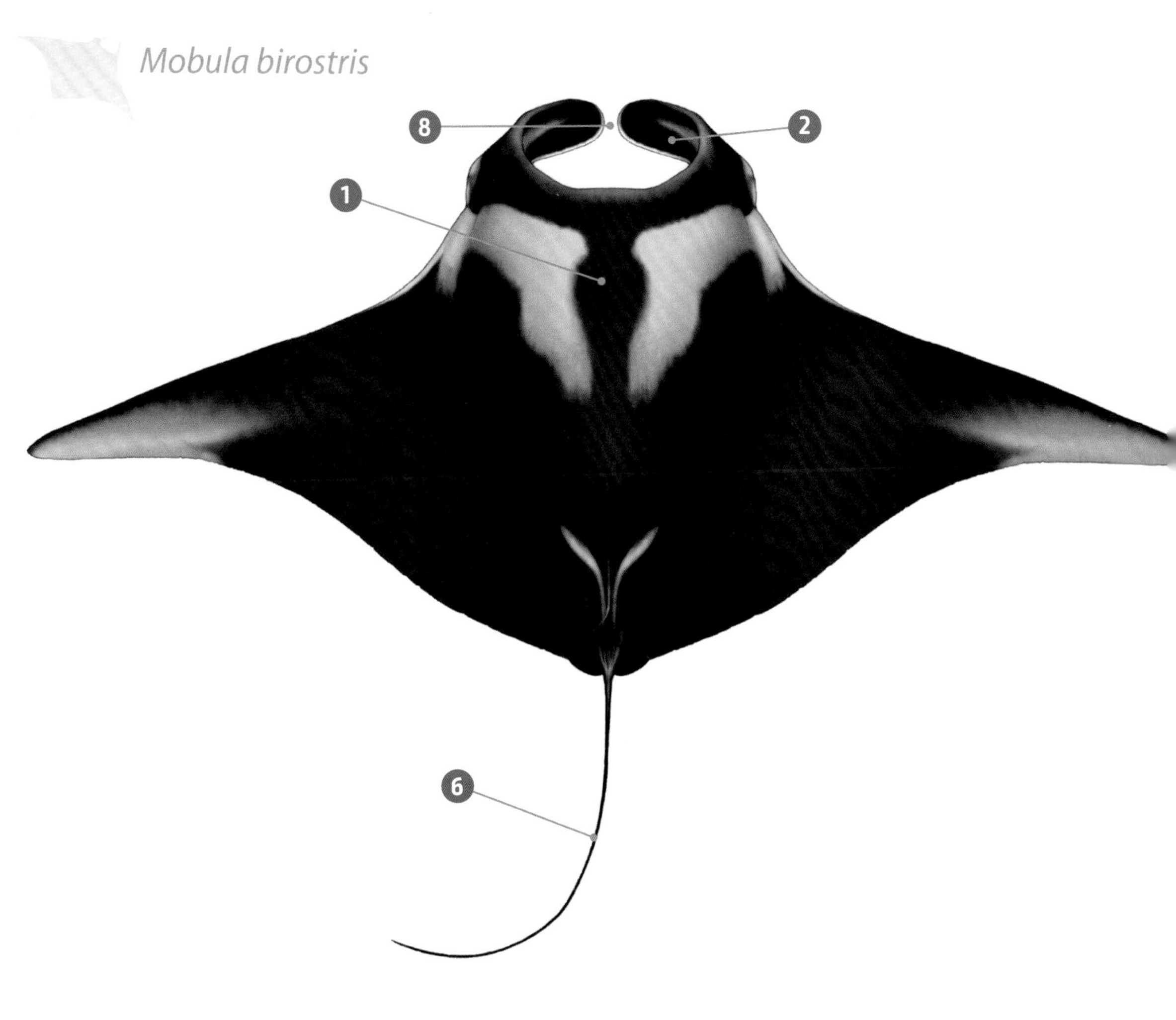

Key features

1. Dorsal white shoulder markings form two mirror image right-angled triangles, creating a 'T' in black.
2. Inside of mouth and cephalic fins, and trailing underside edge of pectoral fins usually shaded black (excluding leucistic morph).
3. If present, ventral spots clustered around lower abdominal region.
4. Gill covers (particularly 5th gill) usually with black shading/flaring.
5. Uniform dorsal fin with no white-tip.
6. Tail equal to or shorter than disc width.
7. Knob-like bulge housing vestigial spine at base of tail.
8. Cephalic fins large, unfurling to meet together in centre of mouth.
9. Spiracle is small transversal slit partially obscured under a flap of skin, above margin of pectoral fin near where it meets body.
10. Large gill plates with fused lateral lobes and a rounded terminal lobe. Plates coloured uniformly black, although occasionally completely white.

Mobula birostris

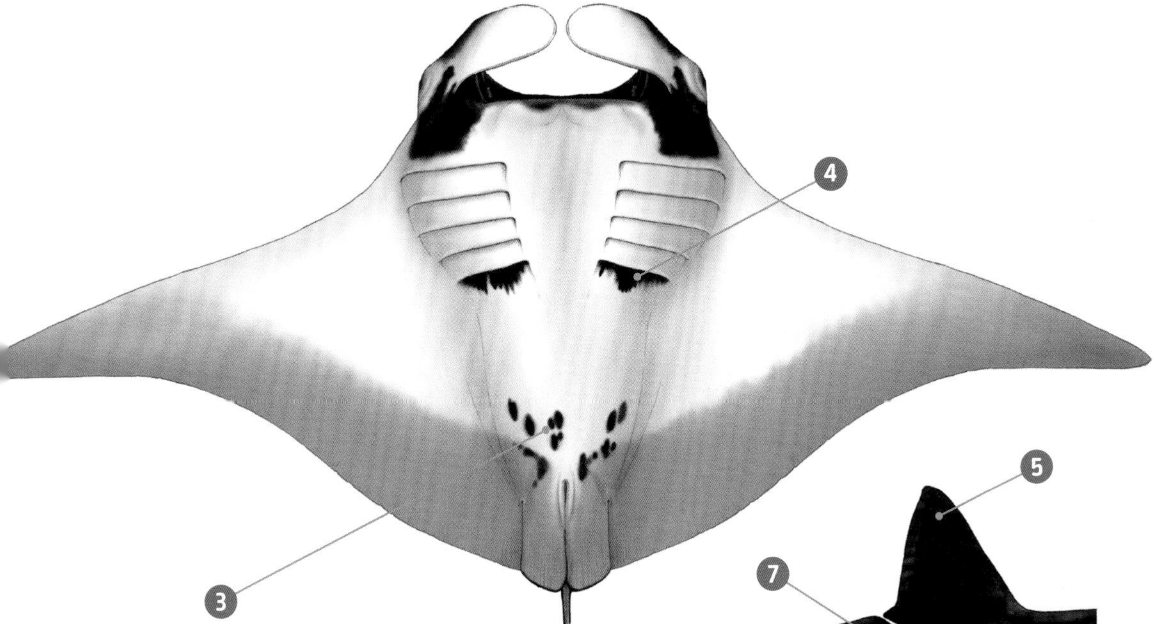
4
3

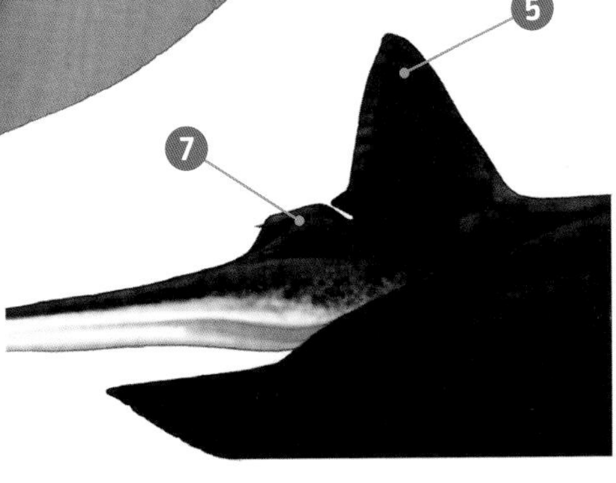
5
7

Lateral view of tail region

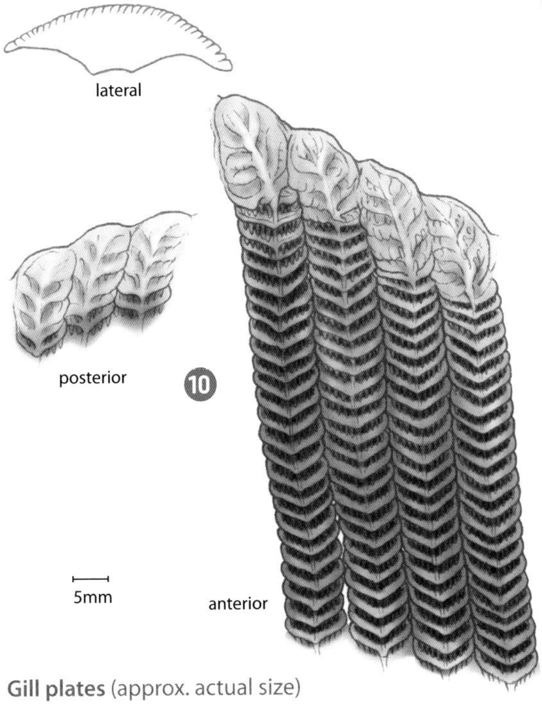
lateral
posterior
10
5mm
anterior

Gill plates (approx. actual size)

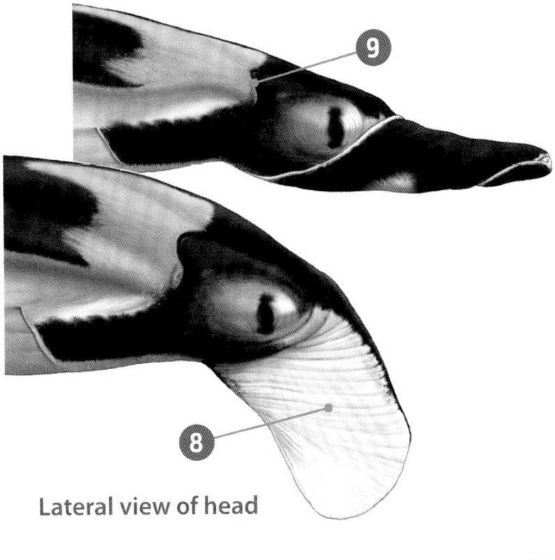
9
8

Lateral view of head

Longhorned Pygmy Devil Ray *Mobula eregoodootenkee* (Bleeker, 1859)

Longhorned Pygmy Devil Ray *Mobula eregoodootenkee*, Raja Ampat, Indonesia.

Species characteristics

Disc width: maximum 130cm (4.3ft), average 110cm (3.6ft).

Weight: up to 25kg (55.1lbs).

Size at maturity: unknown but likely to be around 90–100cm (3–3.3ft).

Age at maturity: unknown.

Lifespan: unknown.

Reproduction: one live-born pup, gestation and frequency is unknown.

Distribution: western Indo-Pacific; east from South Africa, the Red and Arabian Seas, west to eastern Australia, Papua New Guinea and north to Taiwan.

IUCN Red List: **Near Threatened**

FAO species code: **RME**

Longhorned Pygmy Devil Rays *Mobula eregoodootenkee* are small rays, reaching a maximum disc width of just 130cm (4.3ft). The unique gill plate structure, elongate neck and long cephalic lobes make this species easy to identify in fisheries landings. While underwater, the distinct black shading on the leading edge of the pectoral fin at the mid-point ventrally, and the dark brown stripe which runs along the anterior margin of the pectoral fins dorsally, are together useful identifiers. Longhorned Pygmy Devil Rays are found in the Western Indo-Pacific; from eastern South Africa, north into the Red and Arabian Seas, and east across the Indian Ocean to Indonesia and Eastern Australia. Unlike the Shorthorned Pygmy Devil Ray, this species is not reported at remote oceanic islands in the Indian Ocean such as the Maldives, Chagos or Seychelles.

Mobula eregoodootenkee are often observed underwater in schools of a dozen or more individuals as they hunt shoaling bait fish. Swept-back pectoral fins facilitate rapid acceleration, making these extremely fast mini-predators formidable hunters of their fast-moving tropical anchovy and silverside prey.

The primary threats to the Longhorned Pygmy Devil Ray include fisheries throughout their range, where they are

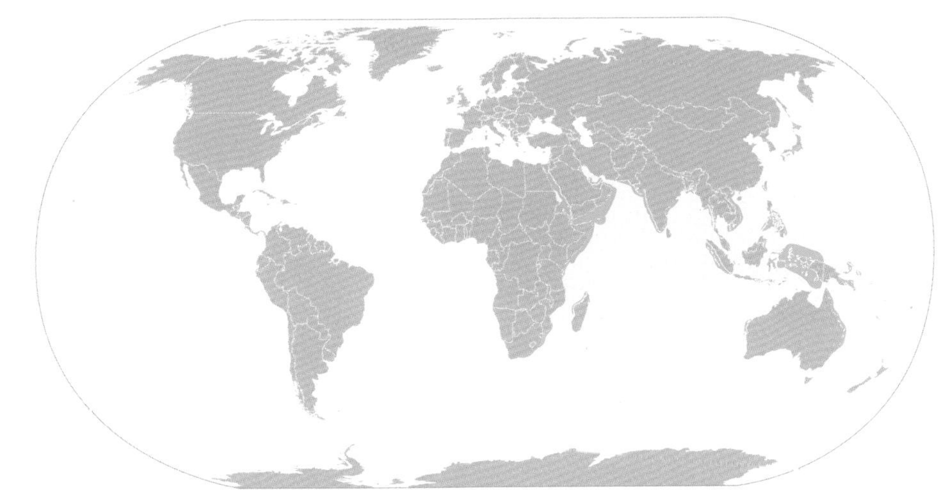

Distribution map of the Longhorned Pygmy Devil Ray *Mobula eregoodootenkee*. Darker areas indicate confirmed range; lighter areas indicate expected range.

landed relatively frequently in Pakistan and Indonesia, but rarely in the Arabian Gulf or throughout India. Being a coastal species, *M. eregoodootenkee* is captured primarily by artisanal fisheries, where data collection is often poor, and although it shares a similar species range as the Shorthorned Pygmy Devil Ray, they have not been captured together in the same net sets or observed underwater in the same habitat by divers. While some research suggests *M. eregoodootenkee* may be a junior synonym of *M. kuhlii*, data from a more comprehensive, expansive and recent genetic study indicates otherwise, supporting the morphological and behavioural variations observed between these two species.

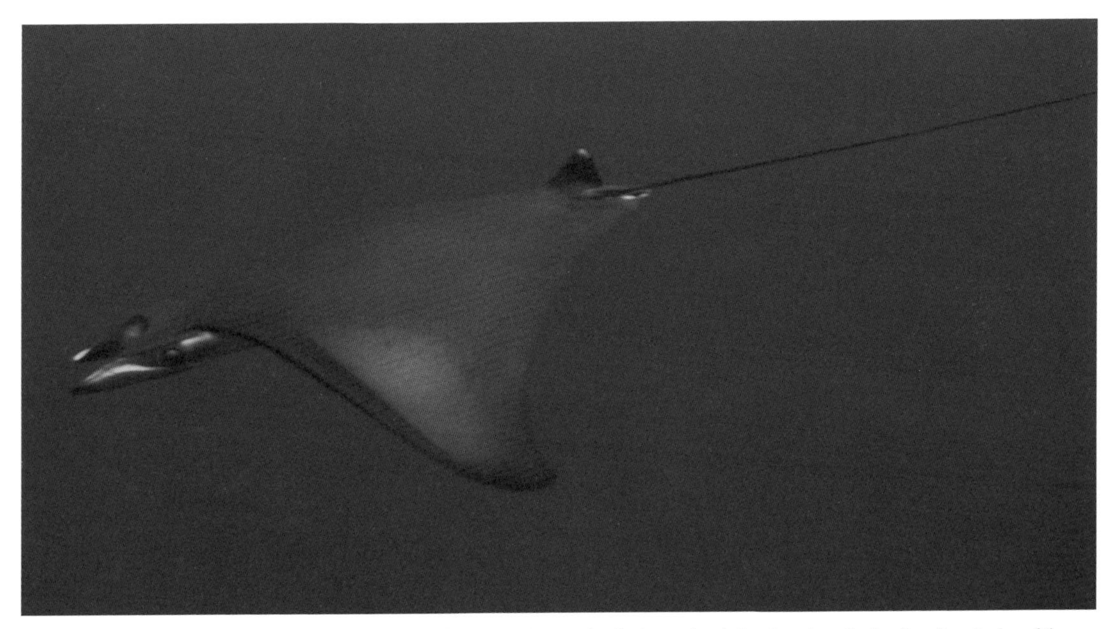

This Longhorned Pygmy Devil Ray *Mobula eregoodootenkee* in Raja Ampat clearly shows the dark stripe along the leading dorsal edge of the pectoral fins, which is a unique identifying feature of this species. All of the other pygmy devil ray species instead have a white leading edge to their pectoral fins dorsally.

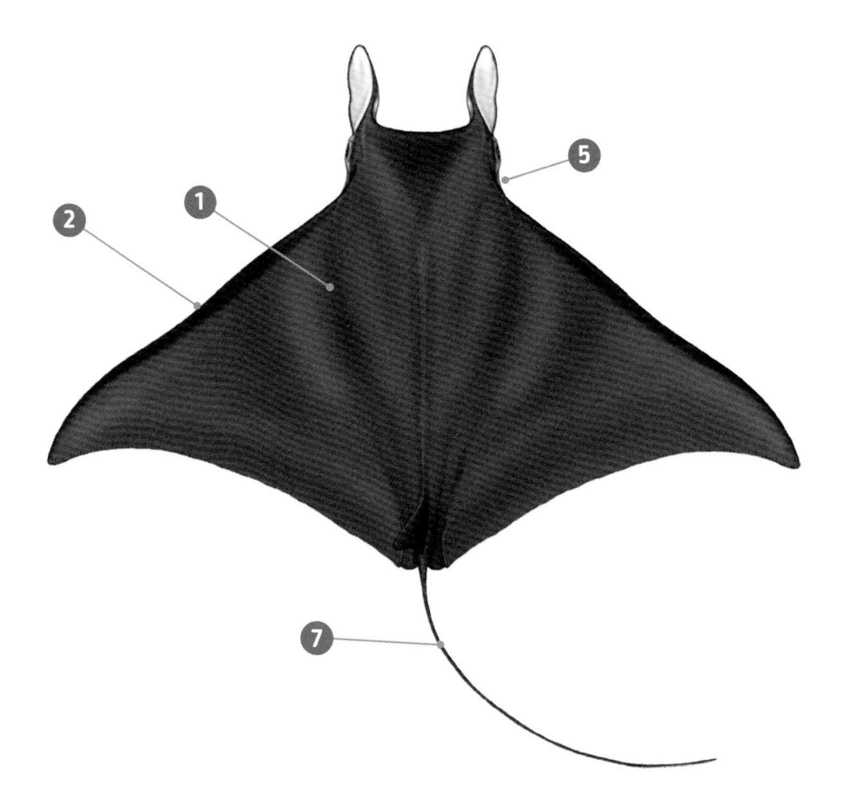

Key features

1. Brown dorsal surface that fades to light grey or black soon after death.
2. Dark brown or grey stripe runs along anterior margin of pectoral fins.
3. White ventral surface with distinct triangular-shaped black to dark-grey shading on leading edge of pectoral fin at the mid-point.
4. Angle of pectoral fins anterior margin sweeps back from head more acutely than in the other pygmy devil ray species.
5. Long-necked appearance.
6. Long cephalic fins; length greater than 16% of total disc width.
7. Tail shorter than disc width. Base of tail dorsally flattened and moderately compressed laterally (quadrangular in section).
8. Often possessing a white-tipped dorsal fin.
9. No spine.
10. Spiracle very small, subcircular and below margin of pectoral fin where it meets body.
11. White ventral markings extend only slightly above eye level.
12. Small to medium-sized gill plates with a highly distinctive structure: few lobes that interlock with those adjacent. Lobes are pinkish-white and leaf-shaped while terminal lobes are very elongated.

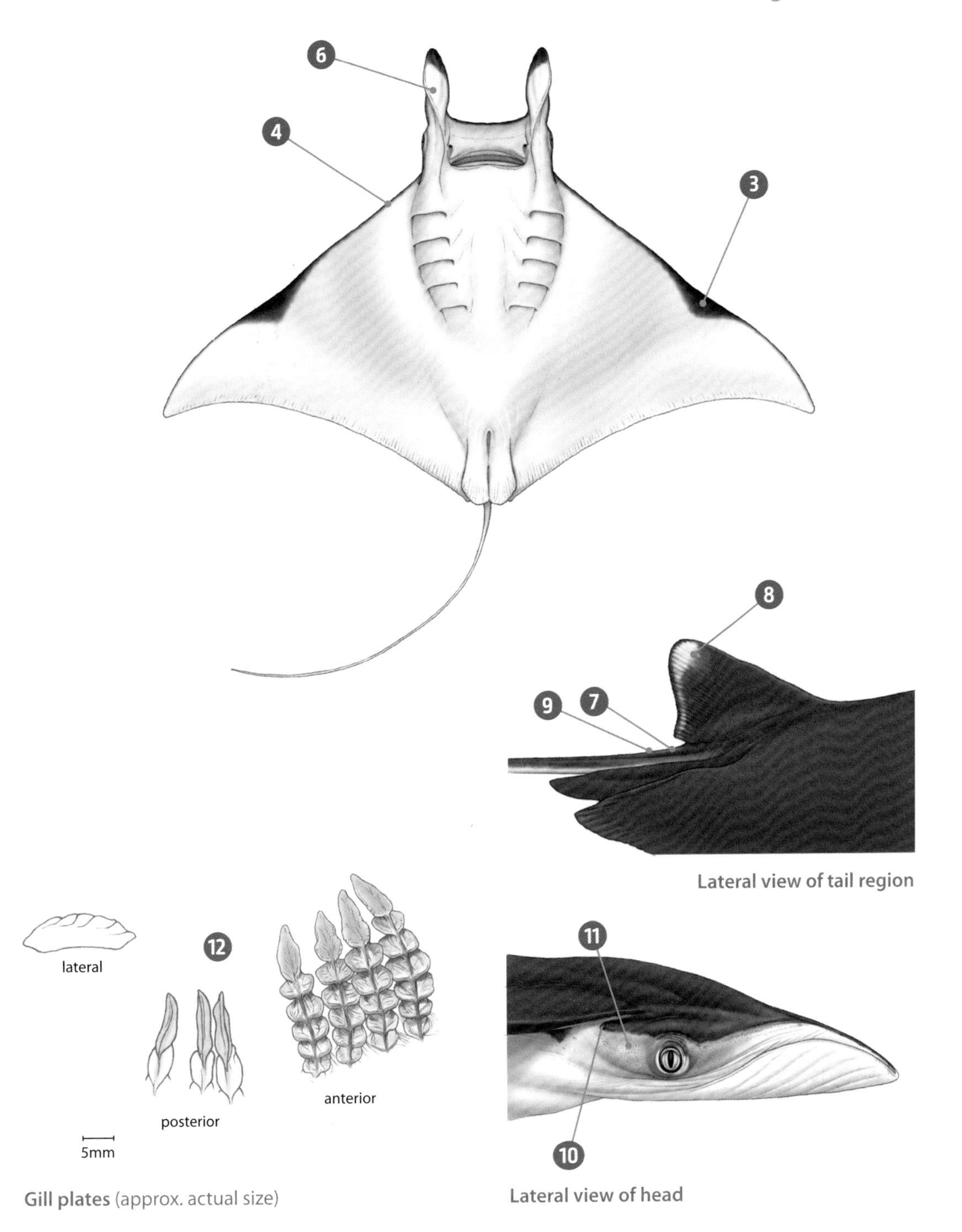

Lateral view of tail region

Lateral view of head

Gill plates (approx. actual size)

West Atlantic Pygmy Devil Ray *Mobula hypostoma* (Bancroft, 1831)

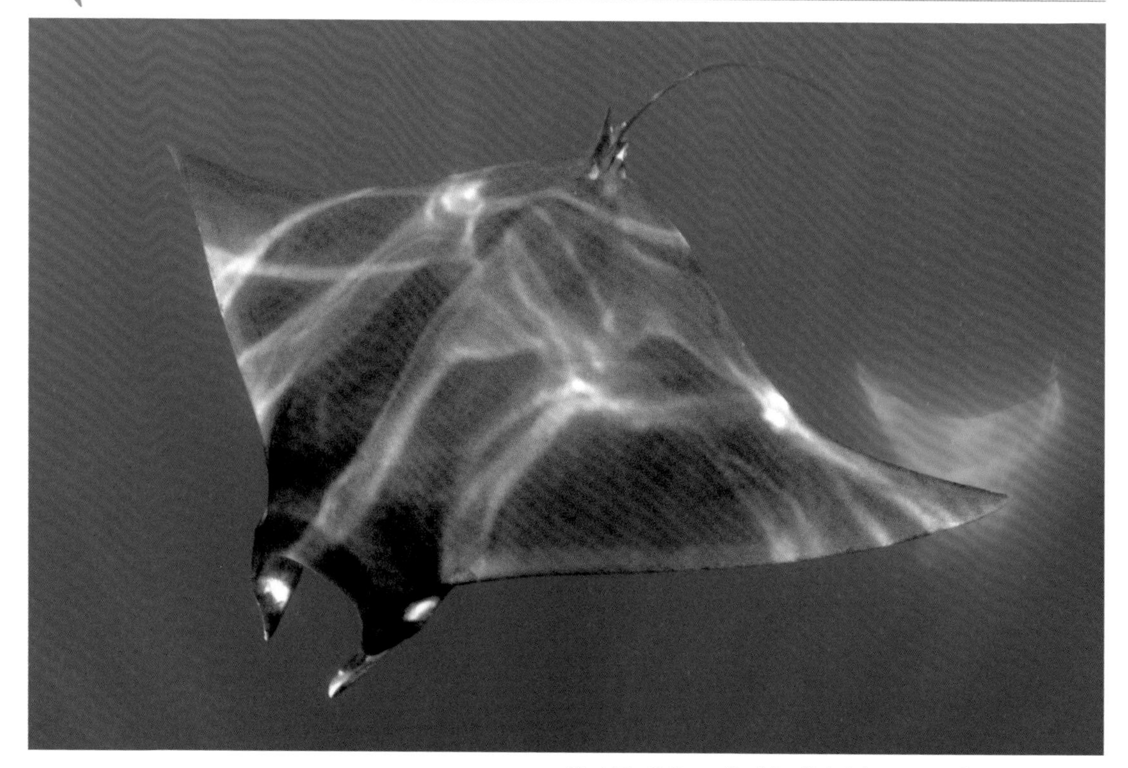

West Atlantic Pygmy Devil Ray *Mobula hypostoma*, Isla Mujeres, Mexico.

Species characteristics

Disc width: maximum 125cm (4.1ft), average 110cm (3.6ft).

Weight: unknown, but likely to be reaching 25kg (55.1lbs).

Size at maturity: ♀ 111cm (3.6ft), ♂ 114cm (3.7ft).

Age at maturity: unknown.

Lifespan: unknown.

Reproduction: one live-born pup; periodicity unknown.

Distribution: coastal waters of the western Atlantic; from North Carolina (USA) to northern Argentina, including the Gulf of Mexico and the Caribbean.

IUCN Red List: **Data Deficient**

FAO species code: **RMH**

West Atlantic Pygmy Devil Rays *Mobula hypostoma* are present only along the Western Atlantic coasts and islands in tropical and warm temperate waters. They are most commonly observed in groups of 2–10 individuals, with groups as large as 40 being recorded. However, it is not unusual to observe solitary individuals.

West Atlantic Pygmy Devil Rays have been frequently observed swimming along shorelines in echelon (line) formation with their mouths positioned right above the sand in extremely shallow water (<1m) (see page 20). With their cephalic fins extended, the feeding devils are likely skimming zooplankton in the surf just off beach breaks.

Attempts at mating by the West Atlantic Pygmy Devil Ray was once observed at the surface, with the female holding her pectoral fins curled up above her back and out of the water to prevent her suitor from biting hold. However, the reproductive behaviour of this species remains poorly documented. Indeed, mating has never been recorded for any of the devil ray species in the wild.

The West Atlantic Pygmy Devil Rays are nowhere directly targeted by fisheries because of their insignificant commercial

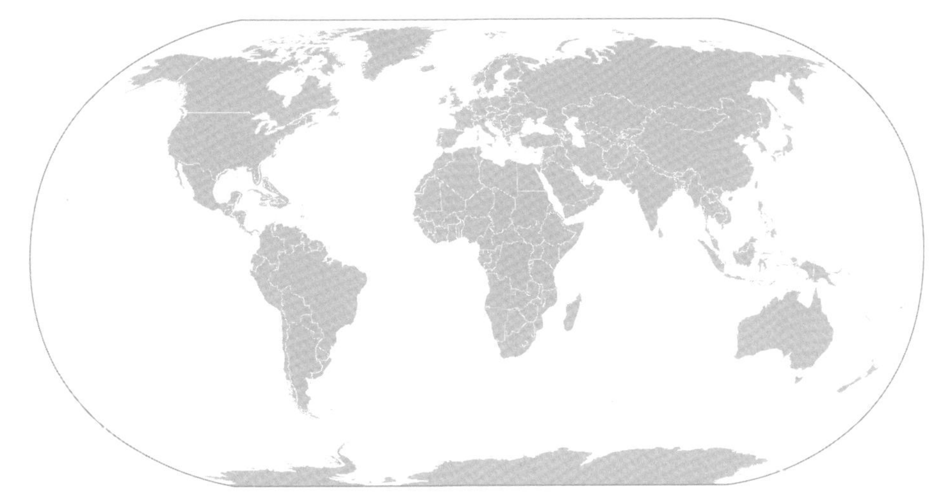

Distribution map of the West Atlantic Pygmy Devil Ray *Mobula hypostoma*. Darker areas indicate confirmed range; lighter areas indicate expected range.

value, however some level of mortality to the species derives from bycatch in longline and net fisheries. Unfortunately, very little information exists on fishery captures of this species throughout its range, which results in its status being defined as Data Deficient on the IUCN's Red List of threatened species.

A school of West Atlantic Pygmy Devil Rays *Mobula hypostoma* off Isla Mujeres, near the tip of the Yucatan Peninsula in Mexico. Observations of these rays can occur in deep, oceanic waters although they seem to be more common in shallower shelf habitats.

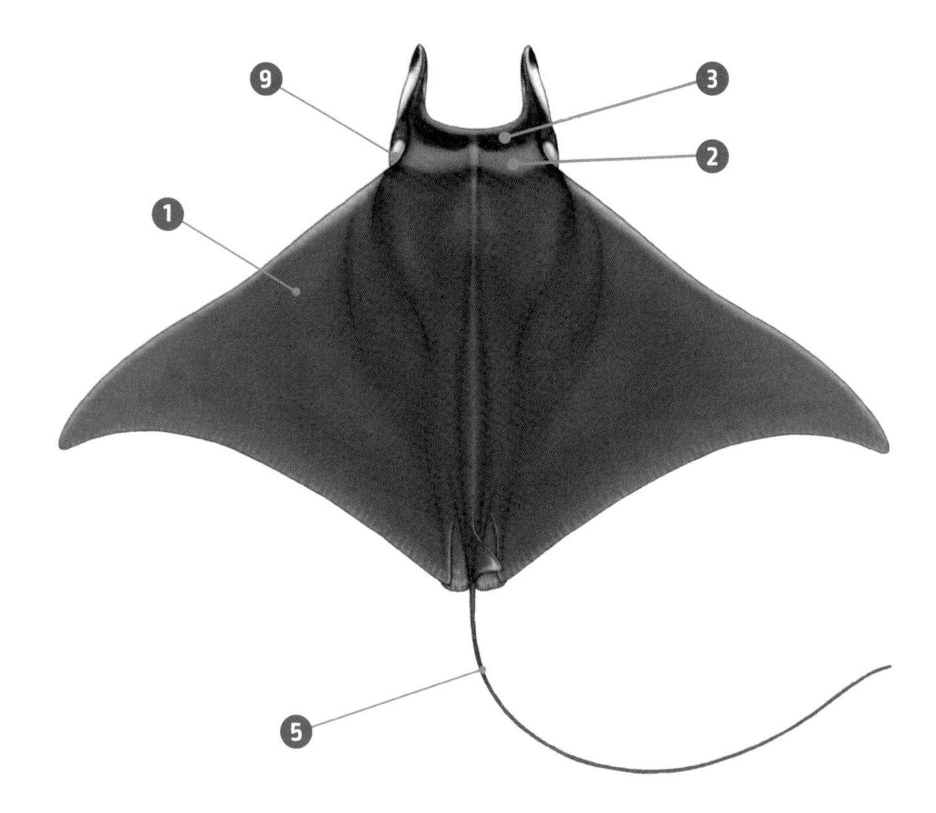

Key features

1. Dorsal surface coloration is variable; from a light brown to grey to black, with some individuals having mottling or spots.
2. Some individuals possess dark grey 'collar', not always connected in the middle, from side to side above spiracles.
3. Most individuals have dark grey stripe running along anterior margin of upper jaw, bordered by a black stripe anteriorly.
4. White ventral surface with light grey on distal ends of pectoral fins.
5. Tail shorter than disc width. Base of tail laterally compressed.
6. No spine.
7. Spiracle very small, subcircular and below margin where pectoral fin meets the body.
8. Dorsal fin with dark rim along along margins, often with a lighter grey area in the middle.
9. White ventral markings wrap up behind and above the eyes, just exceeding margin where pectoral fin joins body.
10. Bronze-brown to grey shading extends ventrally onto anterior of first gill cover near margin where pectoral fin joins body.
11. Small to medium-sized gill plates with leaf-shaped terminal lobe. Plates are grey-white near the base and dark grey-black towards outer lobes, often with lightly coloured terminal lobes. Plates are similar to those of *M. munkiana*.

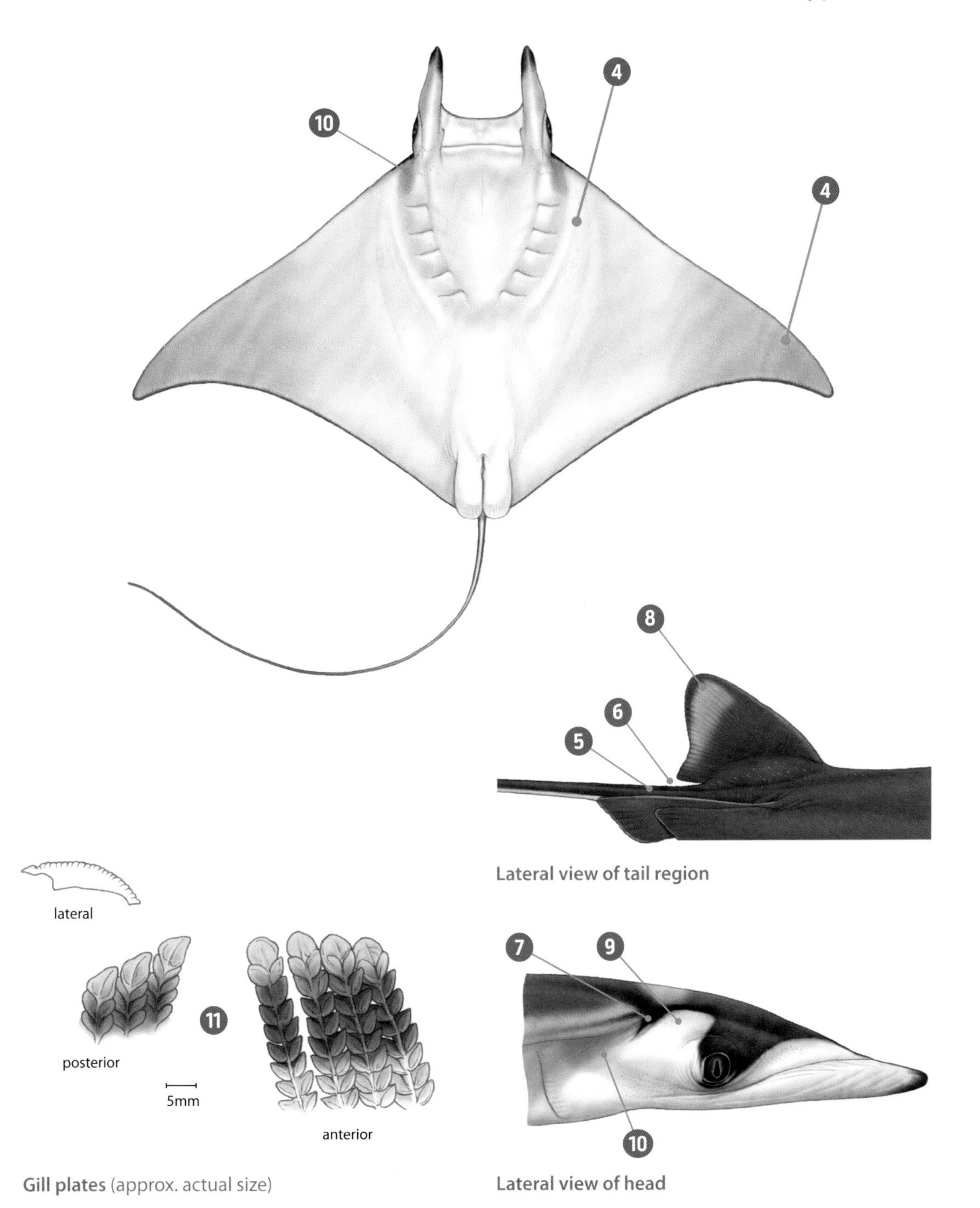

Lateral view of tail region

Gill plates (approx. actual size)

Lateral view of head

Shorthorned Pygmy Devil Ray *Mobula kuhlii* (Müller & Henle, 1841)

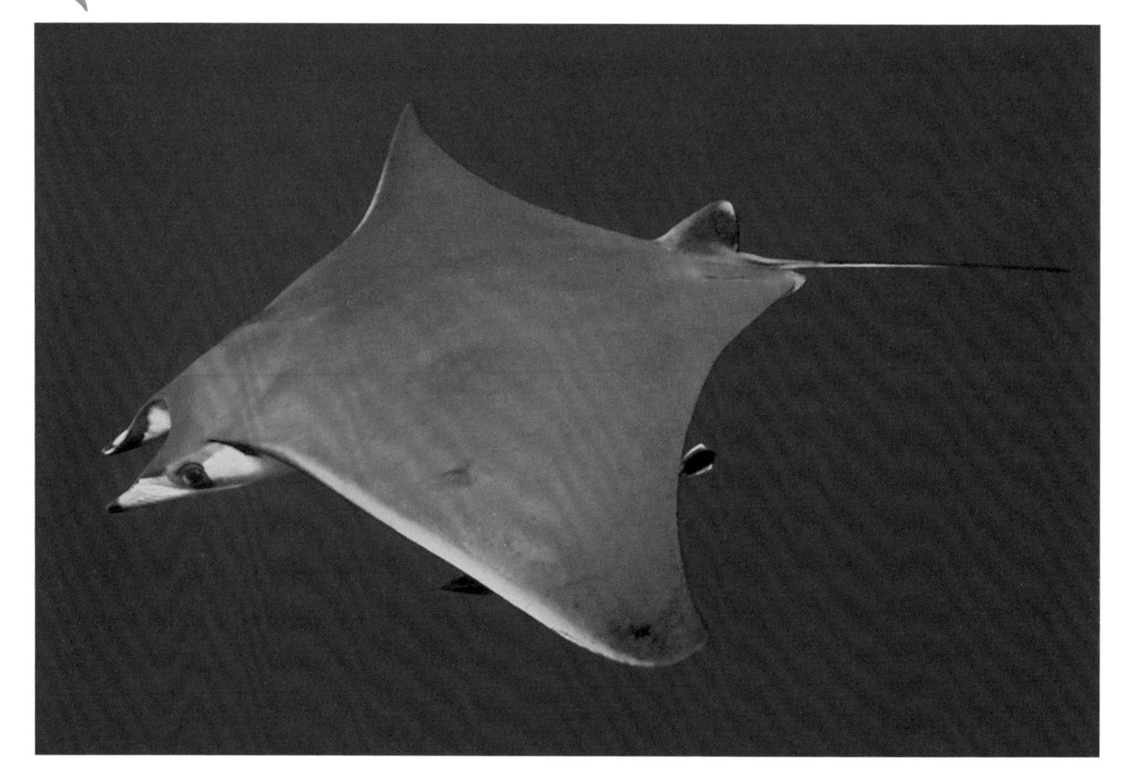

Shorthorned Pygmy Devil Ray *Mobula kuhlii*, Southern Mozambique.

Species characteristics

Disc width: maximum 122cm (4ft), average100cm (3.3 ft).

Weight: up to 25kg (55.1lbs).

Size at maturity: one ♀ was documented pregnant at 116cm (3.8ft), ♂ >103cm (3.4ft).

Age at maturity: unknown.

Lifespan: unknown.

Reproduction: one live-born pup, gestation and frequency is unknown.

Distribution: western Indo-Pacific; from eastern South Africa, the Red Sea, the Arabian Sea, across to Indonesia and northeast Australia.

IUCN Red List: **Data Deficient**

FAO species code: **RMK**

Shorthorned Pygmy Devil Rays *Mobula kuhlii*, with their triangular pectoral fins and relatively short cephalic fins, are extremely agile. These pygmy devils feed individually or in small shoals of 5–15 individuals on swarms of mysid shrimps or larval fish fry. When feeding, they lunge rapidly through the dense schools of their prey with mouths agape. Attacking in small groups appears to confuse their prey, which scatter, presumably making it easier for the devil rays to chase down and catch their meal.

Shorthorned Pygmy Devil Rays have been documented aggregating in schools of 50+ individuals off the coast of Malaysian Borneo and within the atolls of the Maldives. These seasonal aggregations, like those of Munk's Pygmy Devil Rays *Mobula munkiana* in the Sea of Cortez, are thought to be associated with courtship and reproductive activity. In the Maldives these large aggregations peak at the start of the year during the country's northeast monsoon. Some of the females in these schools also appear to be near-term pregnant, and several instances courtship trains have also been recorded.

Shorthorned Pygmy Devil Rays are not frequently encountered in fishery surveys outside of Pakistan, Bangladesh

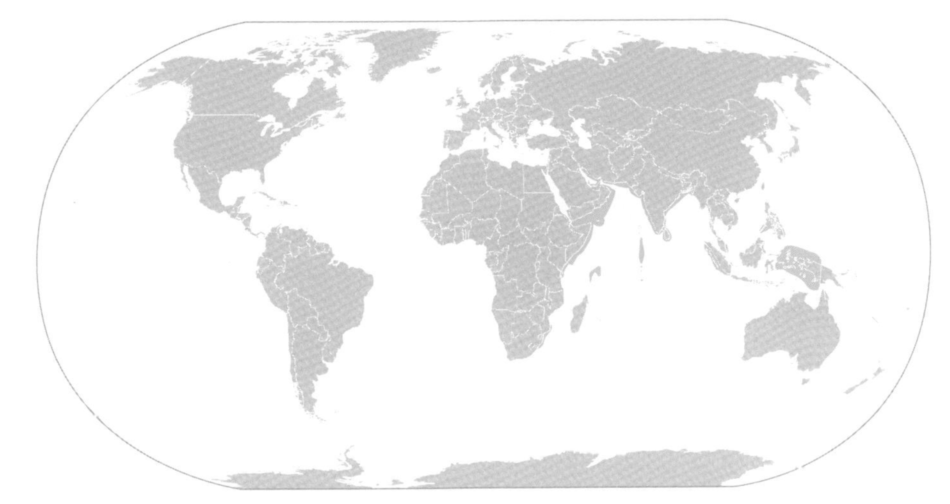

Distribution map of the Shorthorned Pygmy Devil Ray *Mobula kuhlii*. Darker areas indicate confirmed range; lighter areas indicate expected range.

and Mozambique. However, because most survey effort across the range of this nearshore species focuses on pelagic fisheries, sampling bias is likely resulting in the under-reporting of landings by artisanal coastal fisheries. Furthermore, where reporting does occur, dead specimens of these pygmy devils are often misidentified as juvenile Spinetail and Bentfin Devil Rays. Overall, the lack of fisheries data and key life history information for this species throughout its range means its status on the IUCN's Red List of threatened species is defined as Data Deficient.

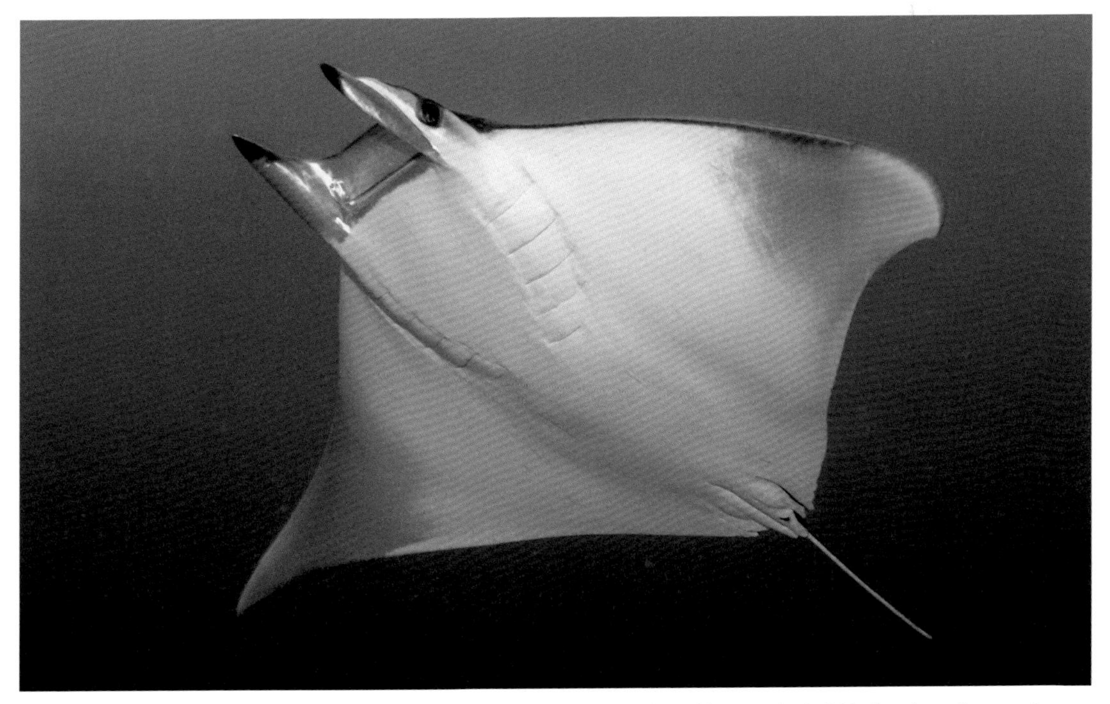

Shorthorned Pygmy Devil Rays *Mobula kuhlii* are commonly encountered by divers in the Maldives cruising individually or in small groups along outer reef crests or inside the channels, like this individual at Lankan Beyru in North Malé Atoll.

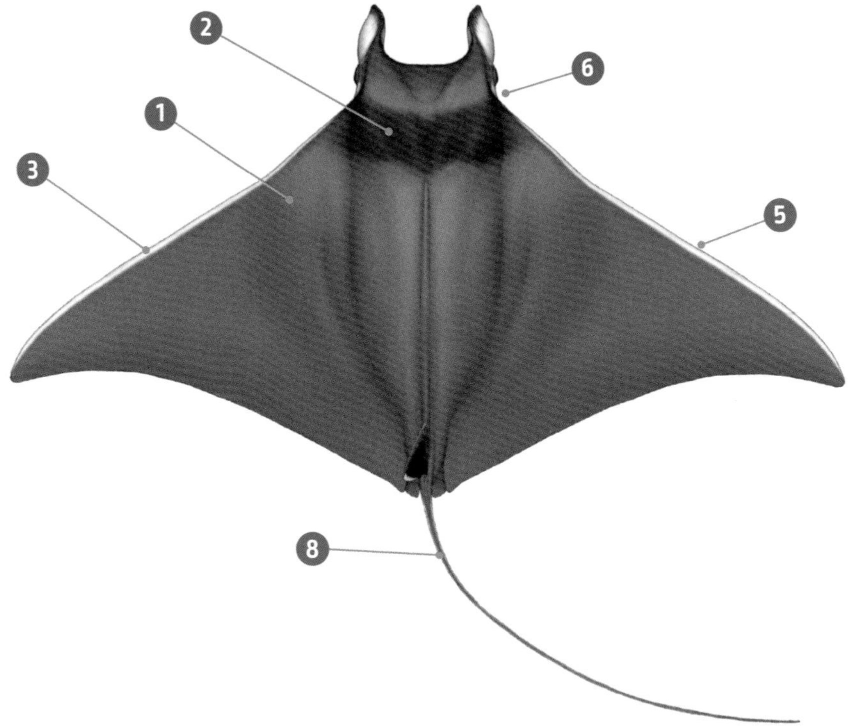

Key features

1. Variable hues dorsally; from mauve, through dark to light grey and chocolate brown. All colours fade to light grey soon after death.
2. Dark 'collar' in half-moon shape from side to side above spiracles.
3. Pale grey stripe runs along the anterior dorsal margin of pectoral fins.
4. Either a complete white ventral surface or with a dark grey-silvery sheen on distal ends of pectoral fins.
5. Anterior margin of pectoral fins smoothly falcate with no angular indentation.
6. Short-necked appearance.
7. Short cephalic fins; length being less than 16% of total disc width.
8. Tail shorter than disc width and counter-shaded throughout. Base of tail dorsally flattened and moderately compressed laterally (quadrangular in section).
9. Often possessing a white-tipped dorsal fin.
10. No spine.
11. Pelvic fin free rear tip extending posteriorly for >50% of dorsal fin base.
12. Spiracle small, subcircular and below margin of pectoral fin where it meets body.
13. White ventral markings do not extend above eye level.
14. Small to medium-sized gill plates with spade-shaped terminal lobe with rounded cusp and distinct median ridge. Plates are grey-white near the base and dark grey-black towards outer lobes with lightly coloured terminal lobes.

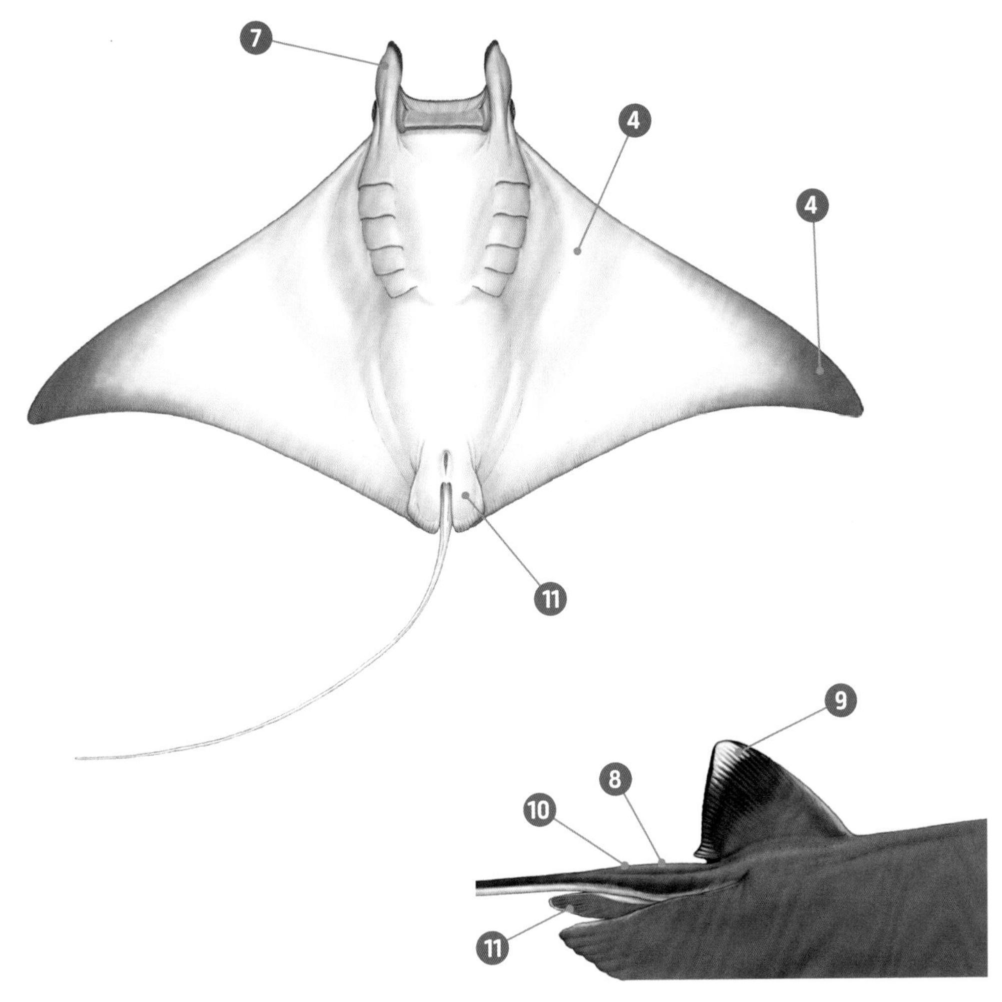

Lateral view of tail region

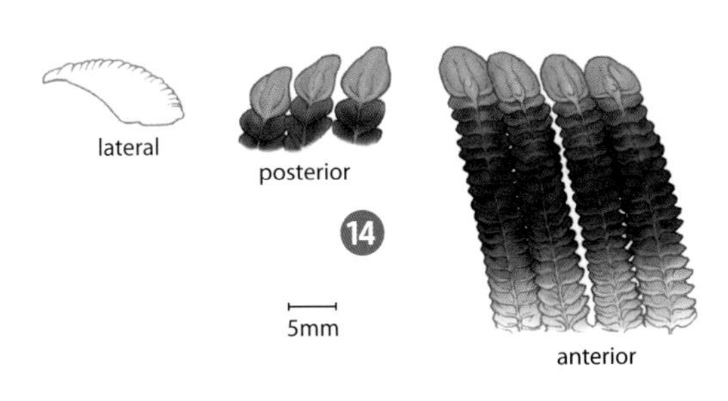

Gill plates (approx. actual size)

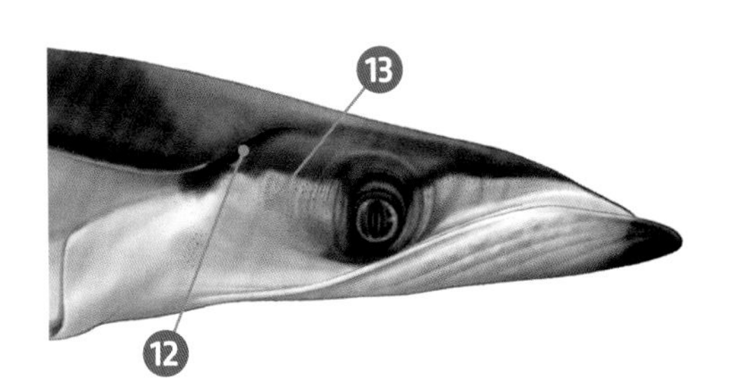

Lateral view of head

Spinetail Devil Ray *Mobula mobular*

(Bonnaterre 1788)

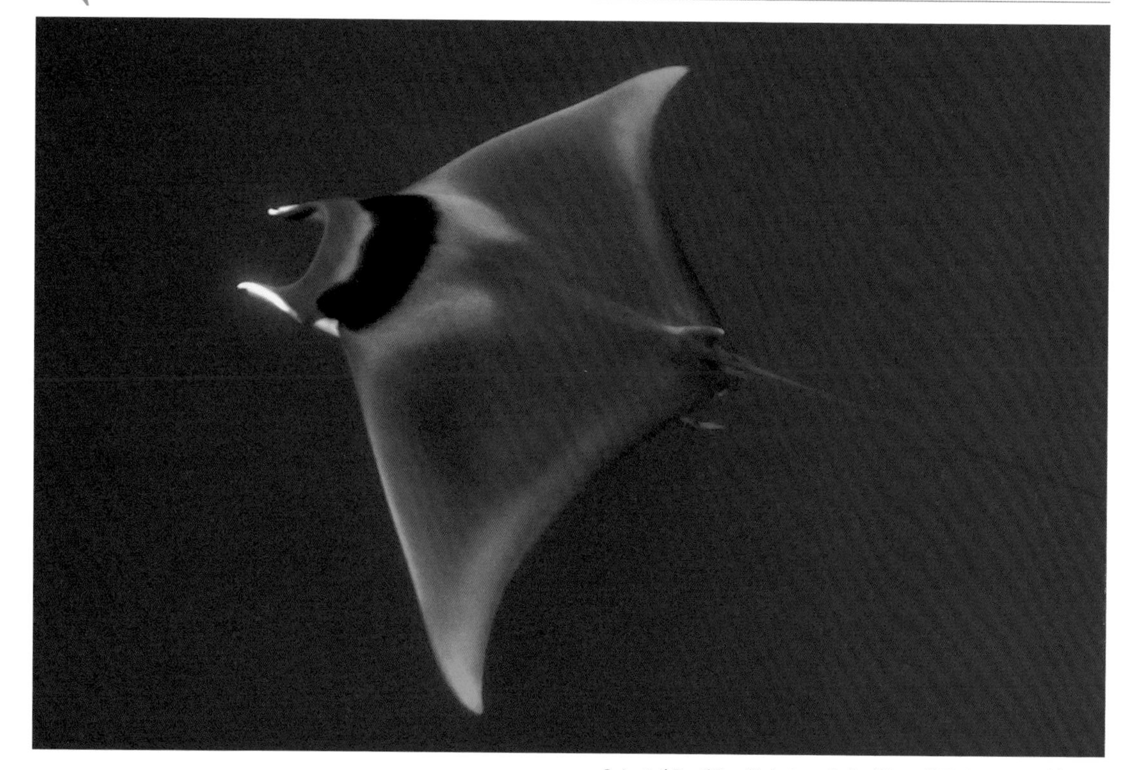

Spinetail Devil Ray *Mobula mobular*, West of Mission Bay, California.

Species characteristics

Disc width: maximum 320cm (10ft), average 180–280cm (5.9–9.2ft).

Weight: up to about 300kg (661lbs).

Size at maturity: ♀ 207cm (6.8ft), ♂ 210cm (6.9ft).

Age at maturity: estimated to be 5–6 years.

Lifespan: unknown, but likely to be at least 15 years.

Reproduction: one pup per litter, periodicity unknown. Size at birth >160cm (5.2ft).

Distribution: circumglobal in tropical and warm temperate waters; including the Mediterranean and Caribbean Seas, the Gulf of Mexico, the Red Sea and the Arabian Gulf.

IUCN Red List: **Near Threatened**

FAO species code: **RMM**

Spinetail Devil Rays *Mobula mobular* are circumtropical, widely diffused in all oceans and adjacent seas, often extending their range into warm temperate latitudes. Until recently the Spinetail Devil Ray was thought to consist of two species: the circumglobal *M. japanica* and *M. mobular*, confined to the Mediterranean Sea. However, recent investigations have revealed sufficient shared genetic and morphological characters to justify the designation of just one species, *M. mobular*, retained because of its taxonomic seniority.

In the Mediterranean Sea the Spinetail Devil Ray has been the subject of recent ecological investigations which revealed its wide distribution across the region; migrating to the northern region of the Sea in summer, and the southeast during the winter. Seasonal migratory behaviour in this species has also been observed in the Gulf of California, where this devil ray feeds predominantly on small planktonic euphausiids (krill).

Inhabiting offshore habitats, Spinetail Devil Rays are among the most commonly caught mobulid rays in fisheries throughout their range. In the Indo-Pacific directed catches are known to occur in India, Sri Lanka, Philippines, Taiwan, Mexico and Peru.

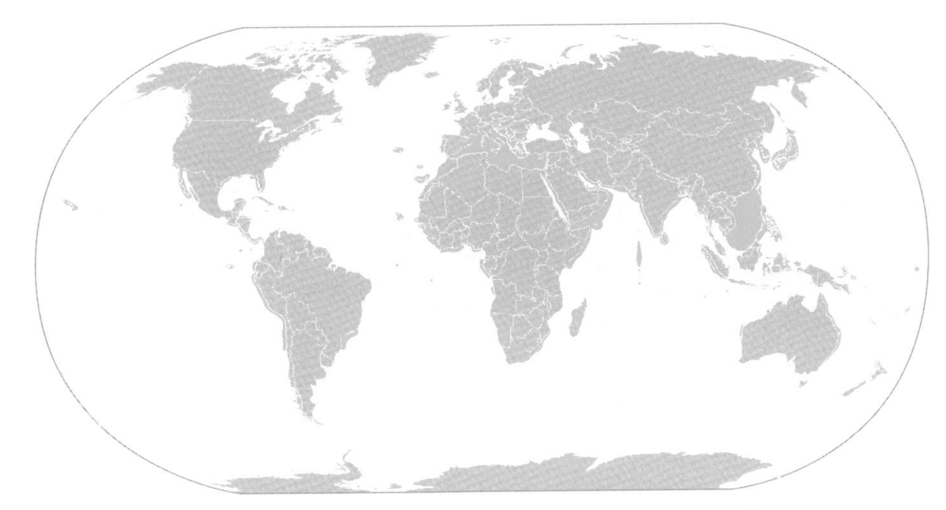

Distribution map of the Spinetail Devil Ray *Mobula mobular*. Darker areas indicate confirmed range; lighter areas indicate expected range.

In the Mediterranean this species is more protected, with the exception of Palestine, where an opportunistic, strongly seasonal (late-winter) target fishery exists. Consequently, given the growing demand in international trade for the dried gill plates of mobulids, Spinetail Devil Rays have been greatly affected, and population declines have been documented throughout most of their range. Their frequent schooling behaviour can also result in entire groups of up to 50 individuals being caught in a single net, further exacerbating their predicament.

Spinetail Devil Rays *Mobula mobular* in the Mediterranean Sea migrate seasonally, presumably to take advantage of seasonal changes in the abundance of their planktonic prey. In this image two males follow a near-term pregnant female.

Key features

1. Thick black band on top of head that stretches from eye to eye, clearly darker than surrounding background colour. Head band is visible only on live individuals.
2. Dorsal surface slate blue, with lighter grey colouring surrounding a black head band and fin edges. When dead, entire dorsal surface quickly fades to black.
3. Bright white ventral surface.
4. Tail equal to or longer than its disc width.
5. White-tipped elongated dorsal fin.
6. A caudal spine (often cut off by fishers).
7. Tail is ventrally flattened at the base of the dorsal fin, soon becoming roundish and very thin.
8. Row of small white tubercles running along either side of the tail.
9. Spiracle is a short transversal slit under a distinct ridge, above margin of pectoral fin near where fin meets body.
10. White ventral markings wrap up behind and above eyes, just exceeding margin where pectoral fin joins body, to meet the black, dorsal head band.
11. Medium-sized gill plates with separated, bristled and pointed lobe edging. They consist of 18–25 lobes, the terminal lobe being leaf shaped. Lobes are coloured black with pinkish-white terminal lobe tips.

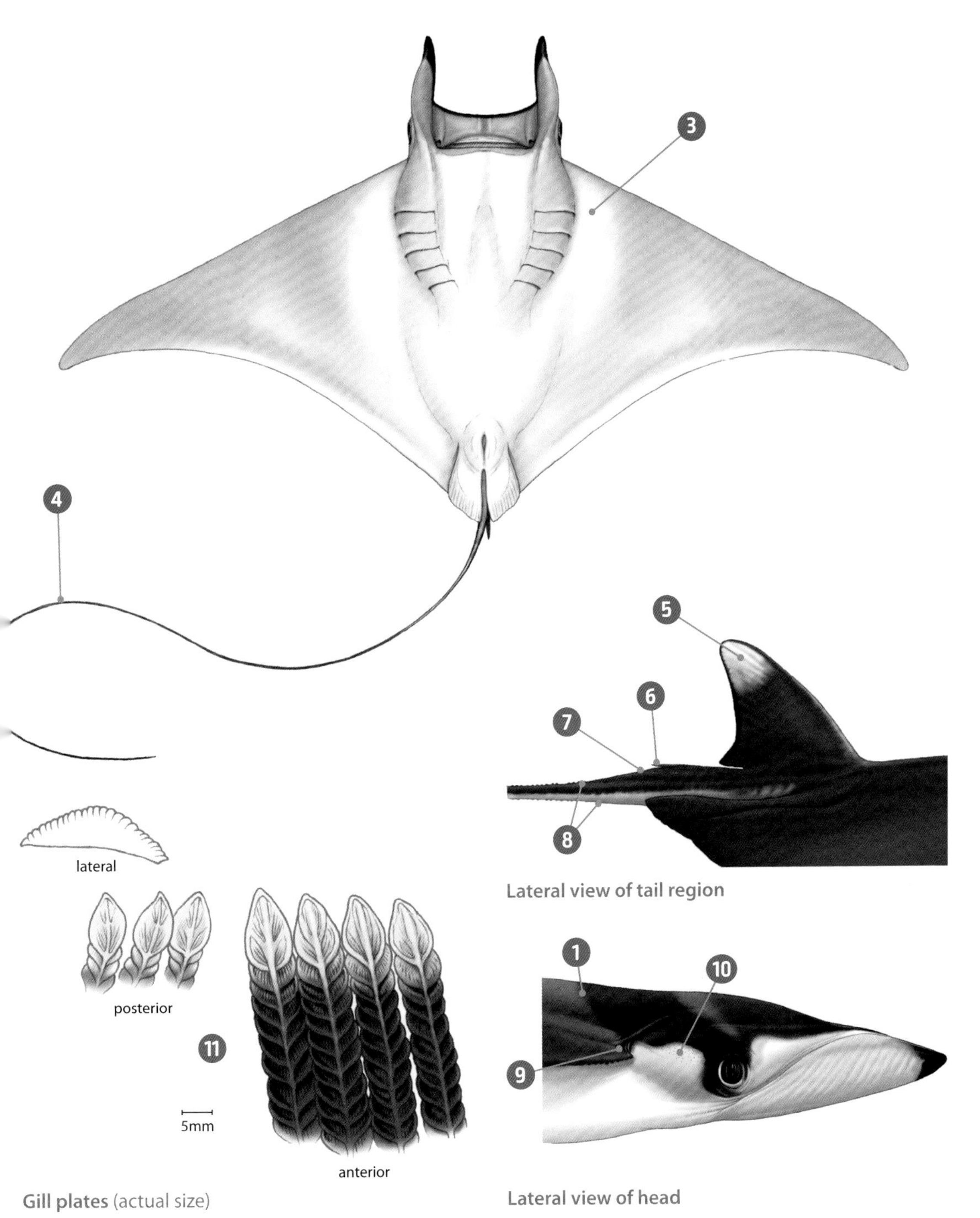

Lateral view of tail region

Lateral view of head

Gill plates (actual size)

Munk's Pygmy Devil Ray *Mobula munkiana*

Notarbartolo di Sciara, 1987

Munk's Pygmy Devil Ray *Mobula munkiana*, Isla Partida, Baja California Sur, Mexico.

Species characteristics

Disc width: maximum 110–130cm (3.6–4.3ft), average 89–100cm (2.9–3.3ft.)

Weight: up to 25kg (55.1lbs).

Size at maturity: ♀ 97cm (3.9ft), ♂ ~87cm (2.9ft).

Age at maturity: unknown.

Lifespan: unknown.

Reproduction: one live-born pup, periodicity unknown.

Distribution: coastal waters of the eastern Tropical Pacific; from Mexico (Sea of Cortez) to Peru, including Galápagos, Cocos and Malpelo islands and adjacent offshore areas.

IUCN Red List: **Near Threatened**

FAO species code: **RMU**

The distribution of Munk's Pygmy Devil Rays *Mobula munkiana* is limited to coastal waters of the Eastern Pacific Ocean from the Sea of Cortez (Mexico) to Peru. The extent of its presence in adjacent offshore areas is unknown. This species is often found in extremely large aggregations of up to tens of thousands of individuals at certain times of the year in the Sea of Cortez. The species' schooling behaviour and migratory movements are likely driven by the ecology of its prey, although no seasonal pattern has been identified yet.

While aggregated, Munk's Pygmy Devil Rays are known to regularly breach, often leaping several metres out of the water in a spectacular display of collective aerial acrobatics. Reasons for this behaviour are still unknown, but possibly linked to social dynamics, particularly as the rays slap their disc down flat on the water's surface upon landing, creating a distinct acoustic pulse.

Munk's Pygmy Devil Rays feed predominantly on small planktonic crustaceans such as mysids. Juveniles have been observed in specific locations of the species' range (e.g., Los Frailes, Mexico), suggesting the existence of specific nursery areas. In these locations the young rays have been observed feeding near the bottom in very shallow waters.

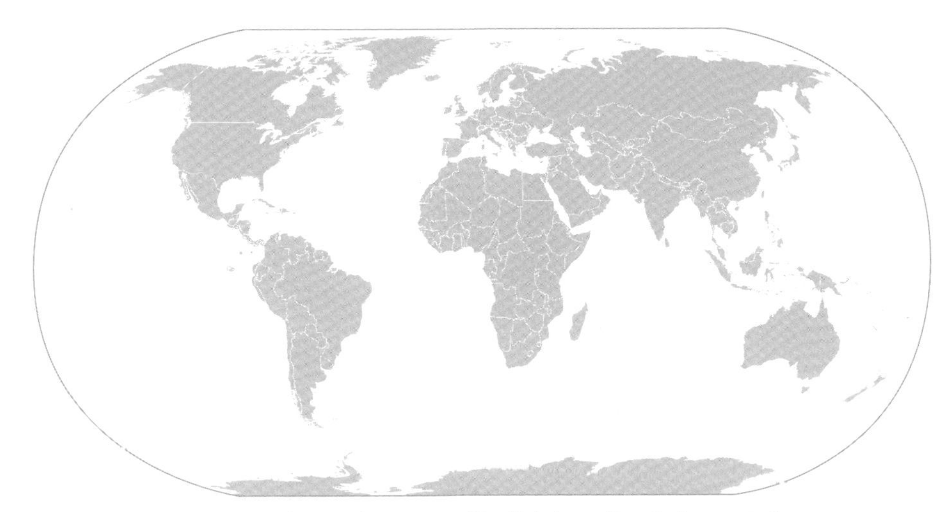

Distribution map of the Munk's Pygmy Devil Ray *Mobula munkiana*. Darker areas indicate confirmed range; lighter areas indicate expected range.

The principal threat to Munk's Pygmy Devil Rays is fishing. These rays are easily caught in both demersal and pelagic gill nets throughout their range, and their typical schooling habits make them particularly vulnerable to such practices. Bycatch in trawl and purse seine is also known to occur and likely to contribute substantially to the species' mortality.

A Munk's Pygmy Devil Ray *Mobula munkiana* leaps several metres out of the water in the Sea of Cortez, Baja California Sur, Mexico. These breaching events regularly occur during mass aggregations of these little devil rays at certain times of the year when thousands of individuals come together, most likely to mate.

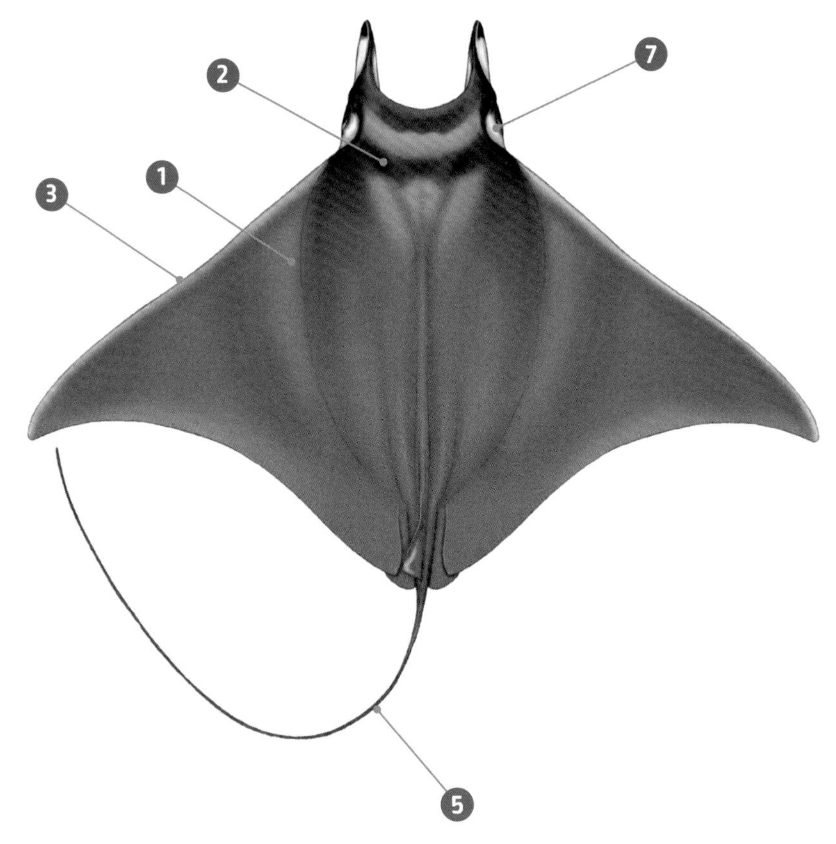

Key features

1. Brownish to mauve-grey dorsally.
2. A dark black head 'collar' is visible in most specimens, with a lighter grey stripe often visible in front, sandwiched between the collar and a dark mouth strip.
3. Light grey stripe runs along anterior dorsal margin of pectoral fins.
4. Whitish ventrally, tending to increasingly dark grey coloration towards distal tips of pectoral fins.
5. Tail shorter than disc width.
6. Dorsal fin with dark rim along margins, often with a lighter grey area in the middle.
7. No caudal spine.
8. Small, round spiracle below margin of the pectoral fin near where it meets the body.
9. White ventral markings wrap up behind and above eyes, just exceeding margin where pectoral fin joins body, to meet the black dorsal head band.
10. Bronze-brown to grey shading extends ventrally onto anterior of first gill cover near margin where pectoral fins join body.
11. Small-sized gill plates, separate from each other. From 9 to 15 ascending lobes per plate, with terminal lobe oval to acorn-shaped. Plates are dark with lighter coloration near base.

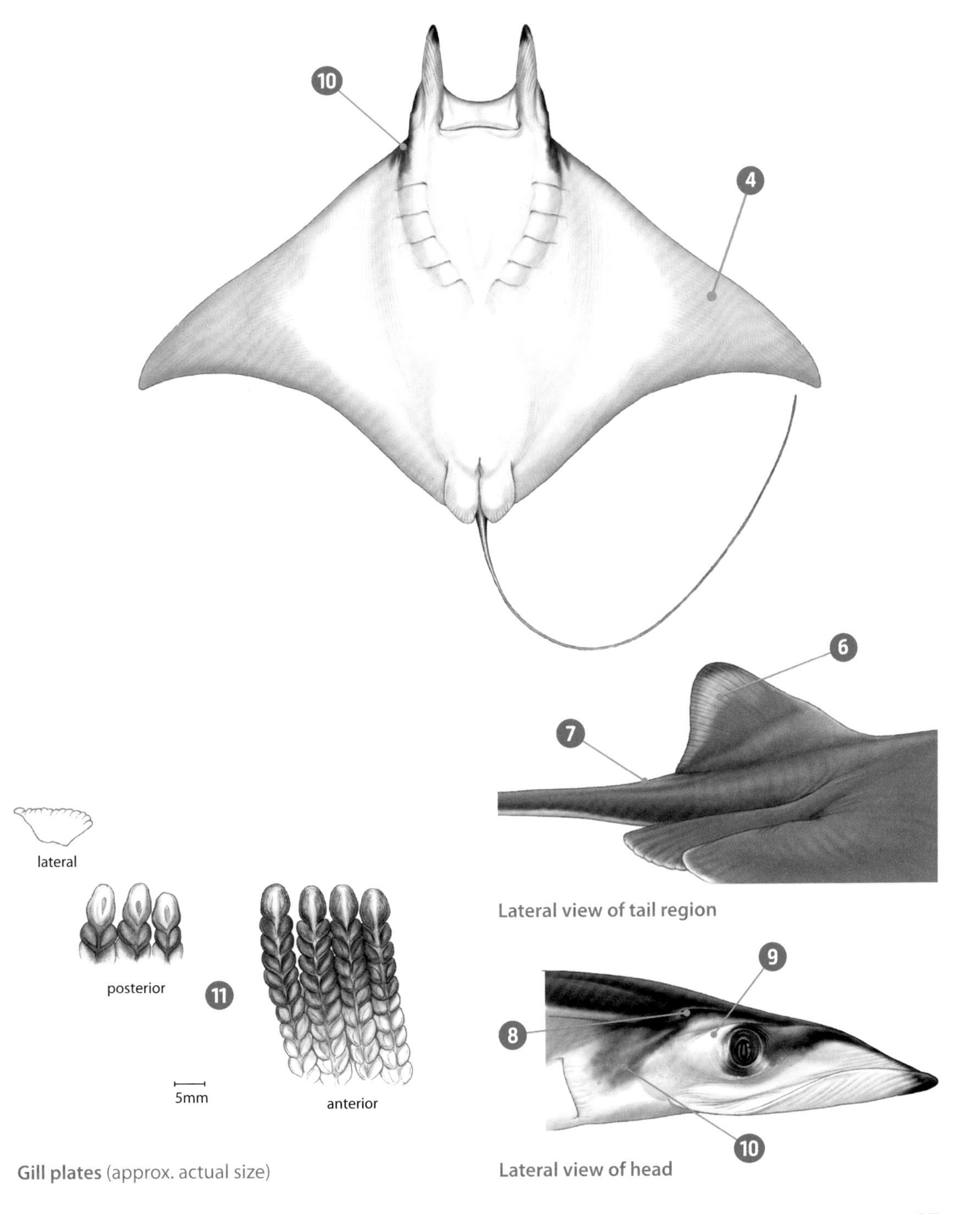

Lateral view of tail region

Lateral view of head

Gill plates (approx. actual size)

East Atlantic Pygmy Devil Ray *Mobula rochebrunei* (Vaillant, 1879)

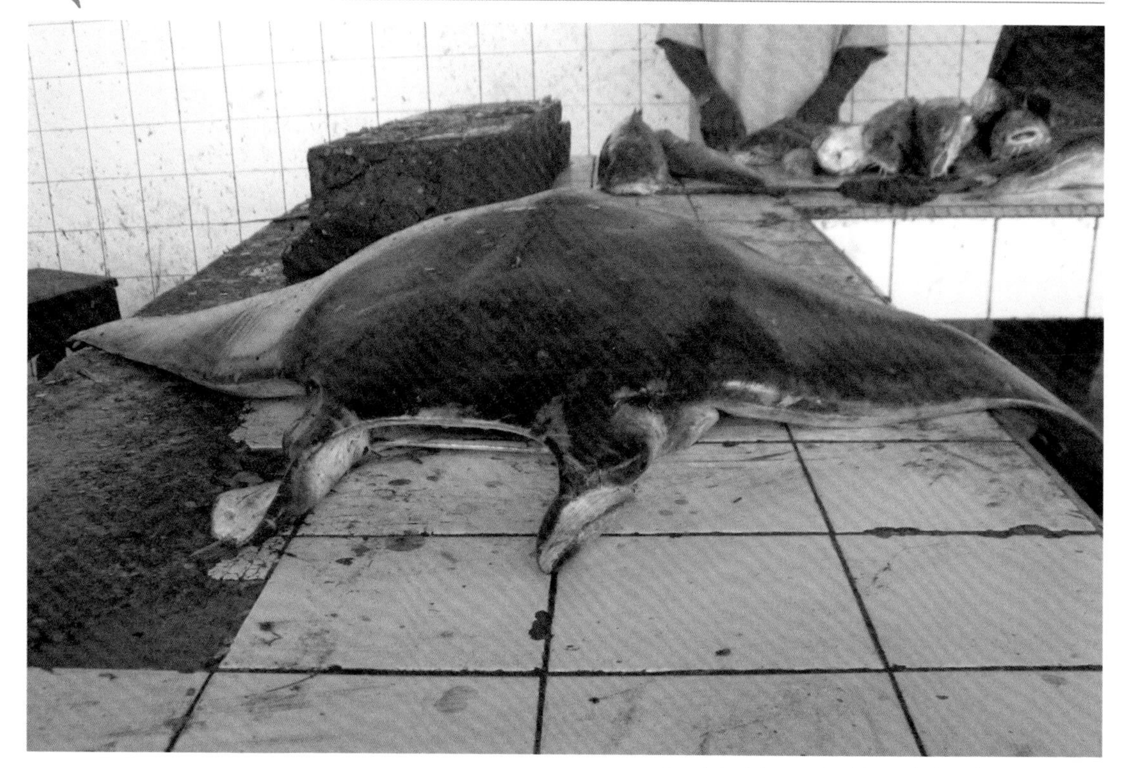

East Atlantic Pygmy Devil Ray *Mobula rochebrunei*, Guinea.

Species characteristics

Disc width: maximum ♀ 133cm (4.4ft), ♂ 125cm (4.1ft), average 113cm (3.7ft).

Weight: up to 30kg (66lbs).

Size at maturity: ♀ ~119cm (3.9ft), ♂ ~110cm (3.6ft).

Age at maturity: unknown.

Lifespan: unknown.

Reproduction: one live-born pup per litter; ~35cm (1ft) at birth.

Distribution: West Africa; Mauritania, Senegal, Guinea, Guinea-Bissau and Angola.

IUCN Red List: **Vulnerable**

FAO species code: **RMN**

The East Atlantic Pygmy Devil Ray *Mobula rochebrunei* occurs pelagically in coastal waters where it travels in schools and feeds on zooplankton and small fish. Like all mobulids, this species is viviparous, giving birth to one young per litter.

Very little is known about this pygmy devil ray, which was last described in some detail in 1960 by J. Cadenat. No verified records of this species have been confirmed since; however, the photograph above, taken in Senegal in 2014, is probably an East Atlantic Pygmy Devil Ray. Unfortunately however, a lack of additional photos or tissue samples for genetic analysis, make verification challenging.

The descriptions of this African pygmy devil ray by Cadenat in 1960 indicate a high similarity to the West Atlantic Pygmy Devil Ray *M. hypostoma*, which is found in the Caribbean Sea and the Western Atlantic Ocean. However, morphometric comparisons by Cadenat revealed that these species were distinct in several ways. Therefore, despite a recent publication suggesting these two mobulids are synonymous and should be considered as a single species, it is the authors' belief that until new specimens of this elusive ray can be attained, and further analysis is undertaken, the species status must remain unchanged.

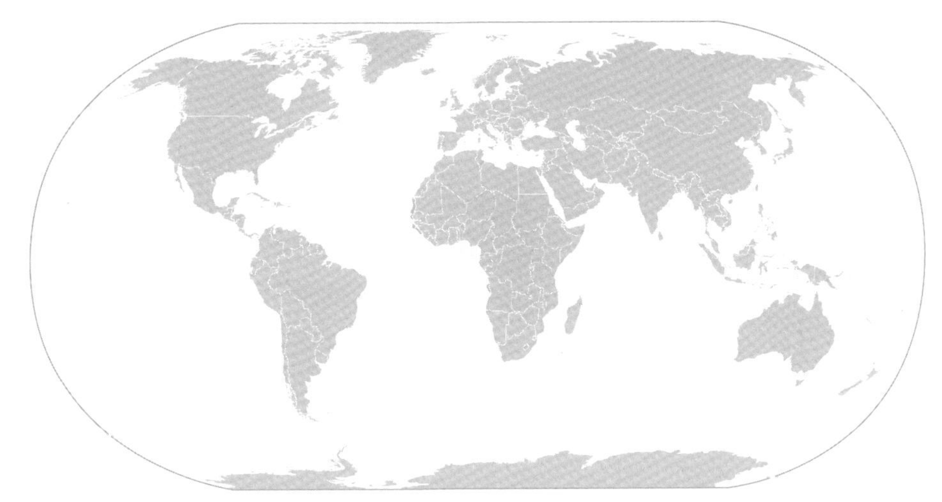

Distribution map of the East Atlantic Pygmy Devil Ray *Mobula rochebrunei*. Darker areas indicate confirmed range; lighter areas indicate expected range.

Despite targeted, but unsuccessful, efforts to collect more specimens for further study in recent years, not enough data is available to calculate accurate population trends for this species. However, given the heavy and unregulated fishing pressure which exists throughout large areas of this species' range, their very low reproductive potential, and their susceptibility to capture, it is highly likely that populations have been heavily impacted. Furthermore, the inability to collect specimens at fish markets throughout several range states where this species was commonly reported in the past, suggests it may already have been fished to near extinction.

Closely related to the East Atlantic Pygmy Devil Ray *Mobula rochebrunei*, the West Atlantic Pygmy Devil Ray *Mobula hypostoma*, like this school captured off Isla Mujeres, in the Yucatan Peninsula, Mexico, also inhabit coastal waters and attain a similar maximum disc width to their East Atlantic relatives.

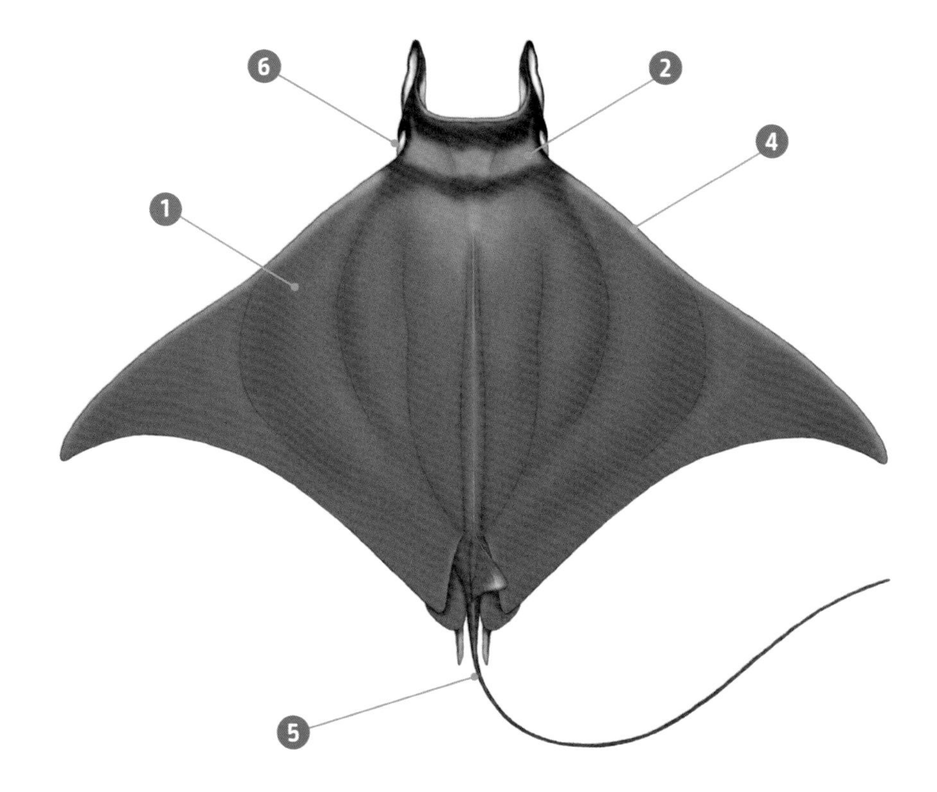

Key features

1. Brownish to mauve-grey dorsally. Colour fading to black when dead.
2. A dark black head 'collar' expected to be visible in live specimens, with a lighter grey stripe in front; sandwiched between the collar and a dark mouth strip. These head patterns fade in all devil rays when dead.
3. Whitish ventrally, tending to an increasingly dark grey coloration towards the distal tips of pectoral fins.
4. Light grey stripe runs along anterior dorsal margin of pectoral fins.
5. Tail shorter than disc width. Base of tail laterally compressed.
6. No spine.
7. Spiracle very small, subcircular and below the margin of pectoral fin where it meets the body.
8. White ventral markings wrap up behind and above eyes, just exceeding margin where pectoral fin joins body.
9. Bronze-brown to grey shading extends ventrally onto anterior of first gill cover near margin where pectoral fins join body.
10. Small to medium-sized gill plates with leaf-shaped terminal lobe. Plates grey-white near base and dark grey-black towards outer lobes. Plates similar to those of *M. munkiana*.

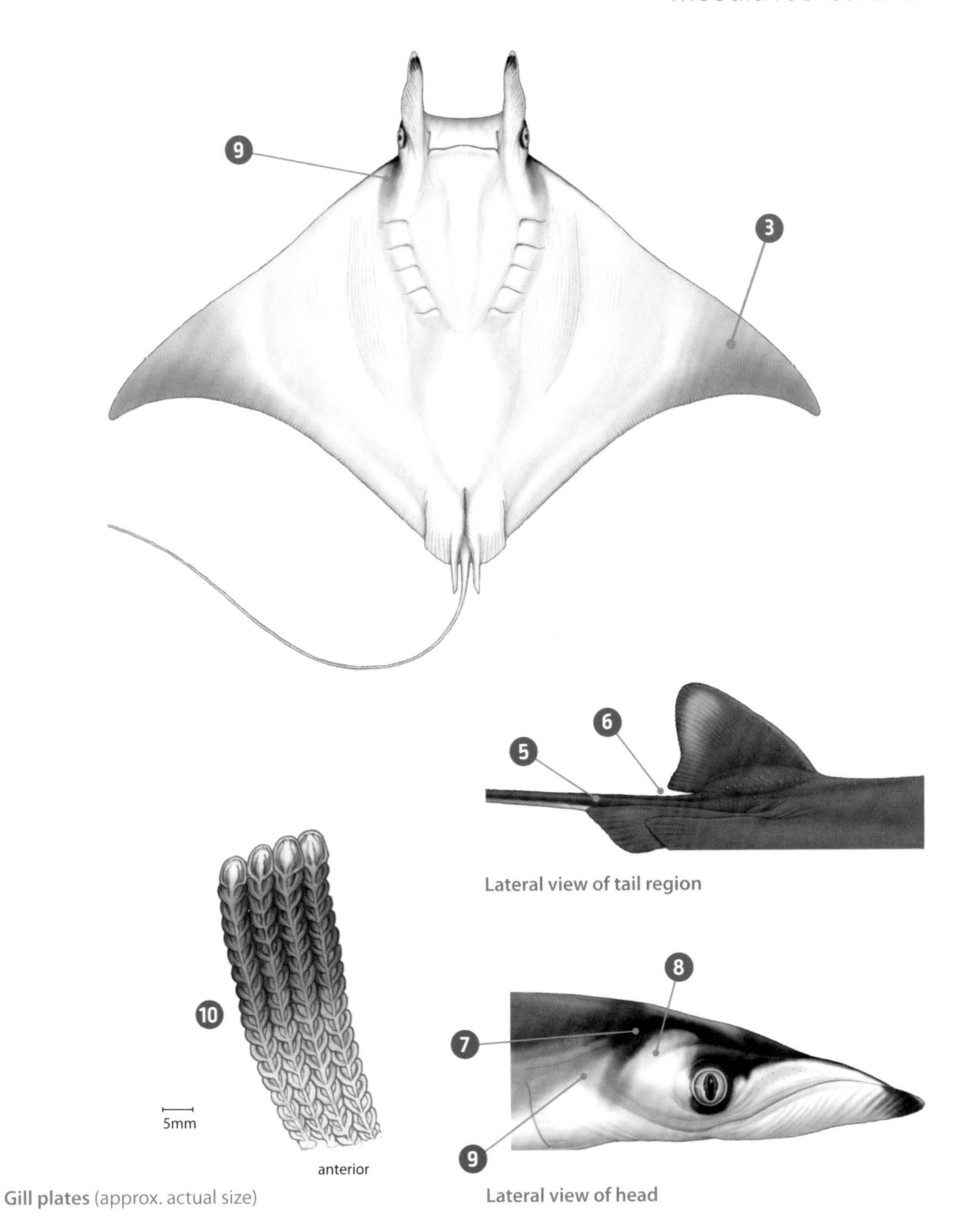

Lateral view of tail region

Gill plates (approx. actual size)

Lateral view of head

Sicklefin Devil Ray *Mobula tarapacana*

(Philippi, 1892)

Sicklefin Devil Ray *Mobula tarapacana*, Princess Alice Bank, Azores, Portugal.

Species characteristics

Disc width: maximum 340cm (11.2ft), average 200–270cm (6.6–8.8ft).

Weight: up to 400kg (882lbs).

Size at maturity: ♀ 270–280cm (8.9–9.2ft), ♂ 240–250cm (7.9–8.2ft).

Age at maturity: unknown, but likely to be at least 5–6 years.

Lifespan: unknown, but likely to be at least 15 years.

Reproduction: one live-born pup per litter; 117–132cm (3.8–4.3ft).

Distribution: found circumtropically in all oceans, to 40°N and 36°S.

IUCN Red List: **Vulnerable**

FAO species code: **RMT**

Sicklefin Devil Rays *Mobula tarapacana* are easily distinguished by their olive-green to brown dorsal surface, and dark ventral markings. However, divers sometimes mistake these species for manta rays (*M. birostris* or *M. alfredi*) due to grey ventral markings. These large rays attain disc widths of 340cm (11.2ft), although most individuals encountered are ~250cm (8.2ft) in disc width.

Sicklefin Devil Rays are a wide-ranging species found throughout the tropics and warm temperate seas, with individuals documented as far north as Japan and as far south as South Africa, Chile and the southern end of New South Wales in Australia. They are also the deepest documented divers among the mobulids, reaching depths of 2,000m (6,562ft) where temperatures plummet to under 4°C, a behaviour indicative of foraging within the deep scattering layers. A sponge-like mesh of large and small arteries in their skull, called a rete mirabile, enables them to keep their brain warm and active during these deep dives. Satellite tagging studies have also revealed that these rays have the ability to travel distances of at least 2,361 miles (3,800km) with speeds of up to 30.4 miles (49km) per day, indicating great potential to cross ocean basins and certainly cross jurisdictional borders.

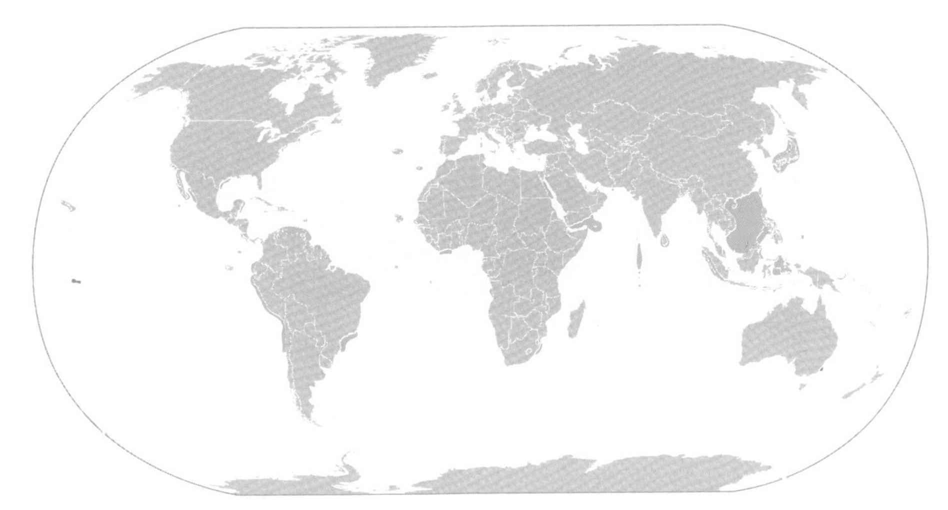

Distribution map of the Sicklefin Devil Ray *Mobula tarapacana*. Darker areas indicate confirmed range; lighter areas indicate expected range.

Sicklefin Devil Rays are considered the 'friendliest' devil rays as they can often be approached by divers, particularly when schooling at offshore pinnacles such as in the Azores. These rays have also been frequently documented shadowing manta rays – a behaviour that is yet to be explained satisfactorily.

The black and white bi-coloration of their gill plates make this species quite identifiable in the gill plate trade, where these 'flower gills' fetch the second highest price after manta gill plates, making this species the third most commonly traded mobulid ray after the Oceanic Manta and Spinetail Devil Ray.

Apart from the two manta ray species, Sicklefin Devil Rays *Mobula tarapacana* are the only mobulids which are regularly accompanied by hitchhiking remoras, like the two stuck to the mouth of this ray off Santa Maria Island in the Azores. The hitchhikers suck onto the rays which such force that they leave behind red sores and permanent lined scars on the skin of the rays (see the lower jaw of the ray above).

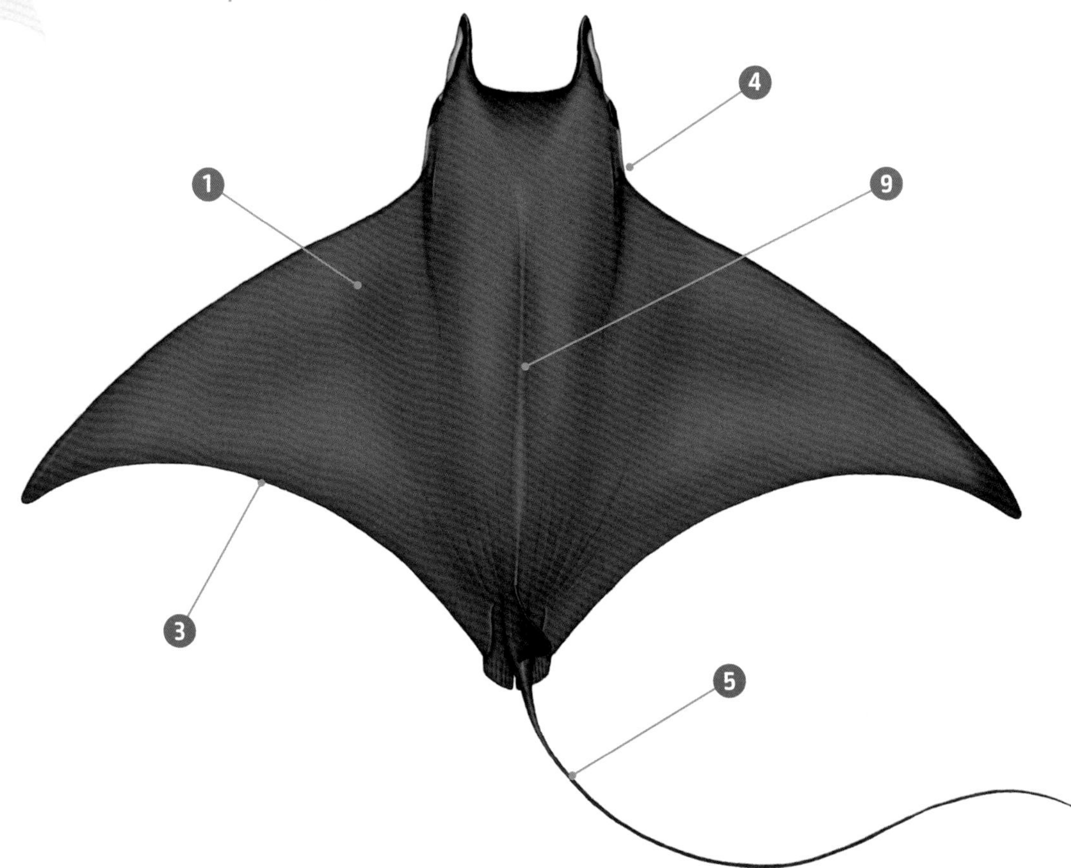

Key features

1. Olive-green or brown dorsal surface.
2. Grey ventral shading on posterior margin of pectoral fins, white anteriorly, with zigzagged margin between both.
3. Trailing edge of pectoral fins distinctly falcate.
4. Long-necked appearance.
5. Tail shorter than its disc width and covered in scales.
6. No spine.
7. Dark grey shading on sides of first gill cover.
8. Plain coloured dorsal fin.
9. Distinctive pronounced ridge along dorsal midline.
10. Grey-white ventral markings do not extend above eye level.
11. Spiracle in an elongated longitudinal slit under a ridge above and behind margin of pectoral fin where it meets body.
12. Medium to large gill plates with fused lobes and rounded terminal lobe with distinctive midline ridge. Plates are distinctively bi-coloured at halfway point with white inner lobes and black outer lobes.

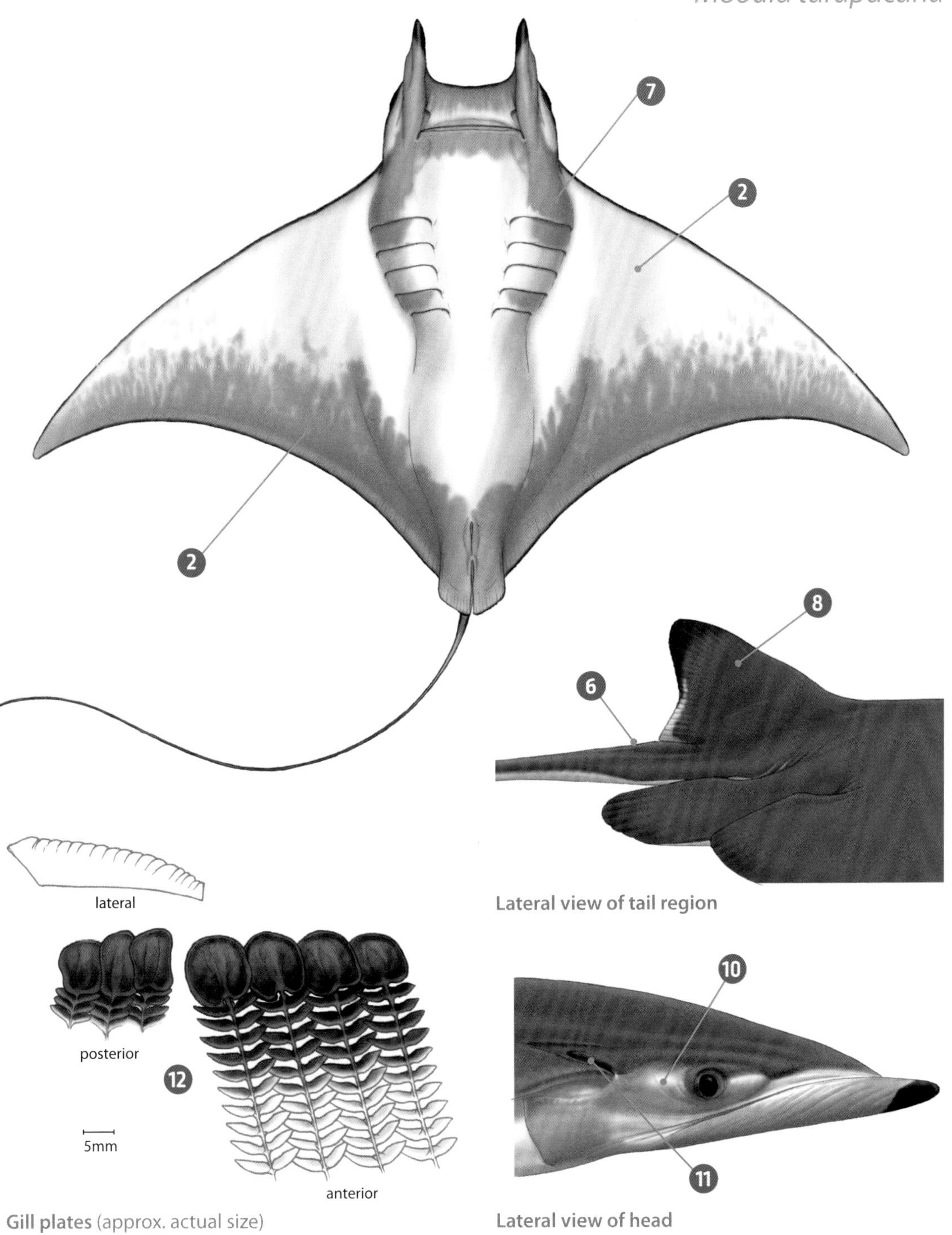

Lateral view of tail region

Gill plates (approx. actual size)

Lateral view of head

Bentfin Devil Ray *Mobula thurstoni*

(Lloyd, 1908)

Bentfin Devil Ray *Mobula thurstoni*, Canary Islands, Spain.

Species characteristics

Disc width: maximum 183cm (6ft),
average 135cm (4.5ft).

Weight: up to 200kg (440lbs).

Size at maturity: ~150cm (4.9ft) for both sexes.

Age at maturity: unknown.

Lifespan: unknown, but estimated to be at least a decade.

Reproduction: one live-born pup per litter; 65–85cm (2.1–2.8ft) at birth.

Distribution: found circumtropically in all oceans; ranging from Japan, Baja California and the Canary Islands in the north, to western Australia, South Africa, Chile and Uruguay in the south.

IUCN Red List: **Near Threatened**

FAO species code: **RMO**

Bentfin Devil Rays *Mobula thurstoni* have a distinctive double curvature on the anterior margin of their pectoral fins, while the undersides of their pectoral fins are coloured a beautiful iridescent silver or gold. This medium-sized devil ray attains a maximum disc width of 183cm (6ft) and has a very long tail, similar to the Spinetail Devil Ray. In the Sea of Cortez, Bentfin Devil Rays have been documented feeding almost exclusively on euphasiid shrimps and occasionally mysid shrimps. Like the other medium and large devil ray species, the Bentfin Devil Ray is an open ocean circumtropical species, coming inshore only occasionally, especially around oceanic islands and seamounts.

Bentfin Devil Rays are usually observed individually, although sometimes small schools of two to six are documented. While rarely captured in fisheries across the Western Indian Ocean, these devil rays are among the most prevalent of mobulids landed across South East Asia and several other locations globally, where the average disc width of the encountered individuals also appears to be slightly larger.

The predominantly black gill plates of Bentfin Devil Rays often have some white shading. However, unlike the distinctive border between the black and white areas on the gill plates of Sicklefin

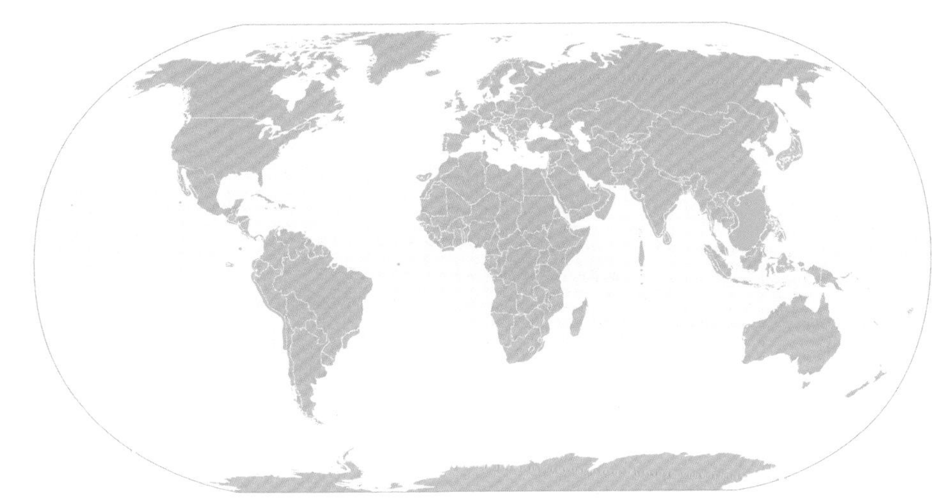

Distribution map of the Bentfin Devil Ray *Mobula thurstoni*. Darker areas indicate confirmed range; lighter areas indicate expected range.

Devil Rays, the shading variation is less delineated in this species. Due to the increasing international trade in mobulid ray gill plates globally, which has led to the expansion of largely unregulated and unmonitored fisheries, coupled with their sensitivity to even moderate levels of fishing pressure as a result of their extremely low reproductive rates, Bentfin Devil Rays were recently listed as Near Threatened on the IUCN's Red List of threatened species.

When observed underwater, the distal portion of the Bentfin Devil Ray's *Mobula thurstoni* pectoral fins ventrally iridesce a beautiful gold or silver when the light is reflected off the fin's surface. This individual was photographed in the waters off Malapascua Island in Cebu, Philippines.

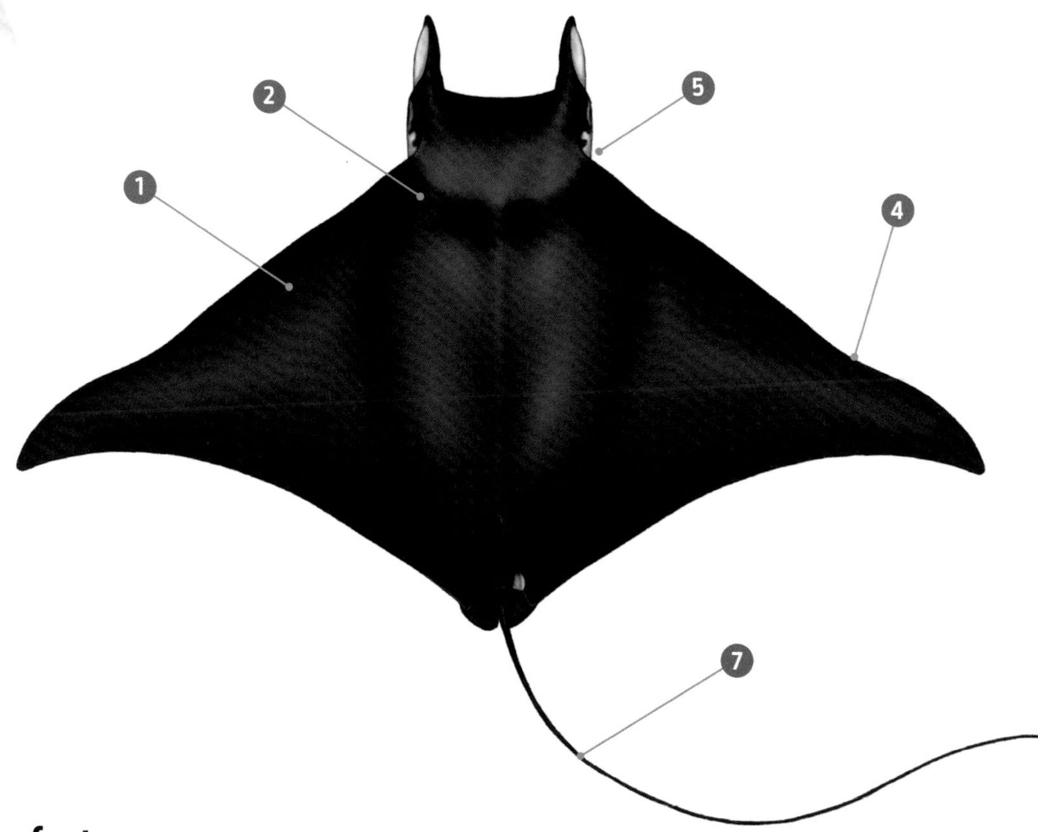

Key features

1. Dark mauve to blue-grey dorsally. Colour fading to black when dead.
2. Thick black band on top of head that stretches from eye to eye, clearly darker than surrounding background colour. This band is only visible on live individuals.
3. White ventral surface with silver-brown sheen on distal ends of pectoral fins.
4. Anterior margin of pectoral fins have a distinctive double curvature with black-grey shading on curve.
5. Short-necked appearance.
6. Short cephalic fins; length being less than 16% of total disc width.
7. Tail equal to or longer than disc width in length when fully intact and dorso-ventrally compressed just behind dorsal fin for about one length of dorsal fin base.
8. White-tipped dorsal fin.
9. No spine.
10. Long pelvic fins that extend behind pectorals by a distance equal to 40% of dorsal fin base.
11. Spiracle small, subcircular and below margin of pectoral fin where it meets body.
12. White ventral markings do not extend above eye level.
13. Small to medium-sized gill plates with leaf-shaped terminal lobe twice as long as wide. Plates black with grey terminal lobes and some light-white shading near base.

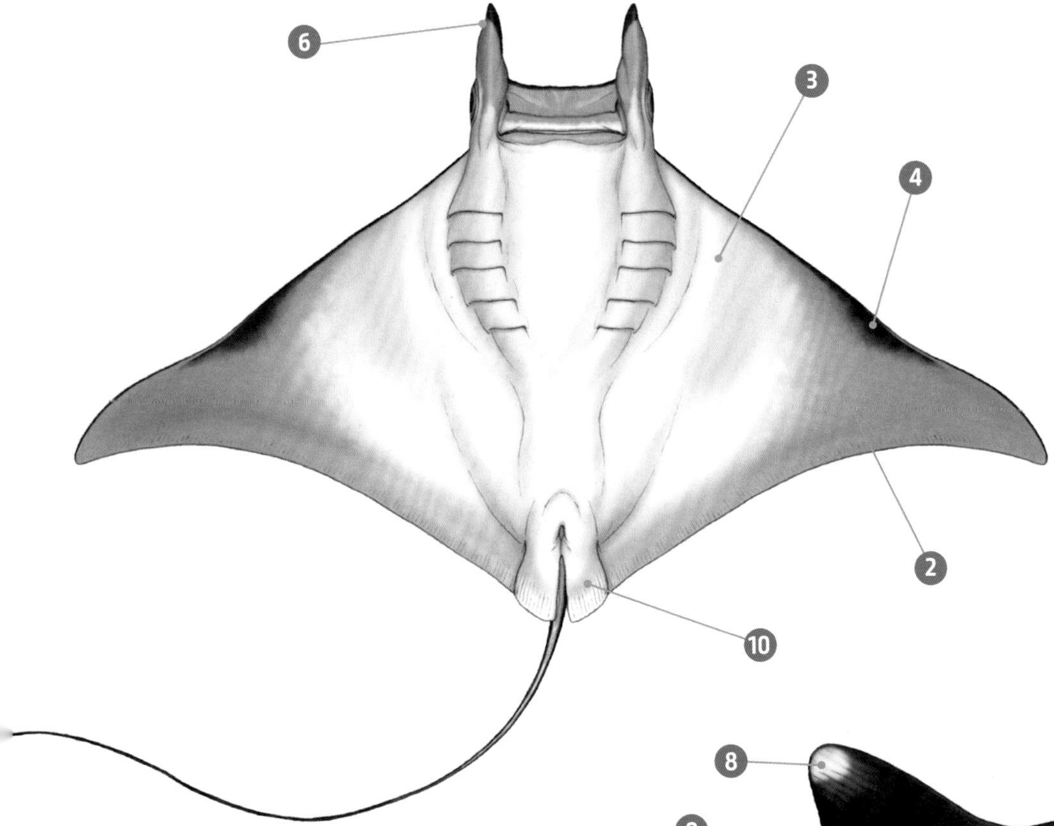

6 3 4 2 10

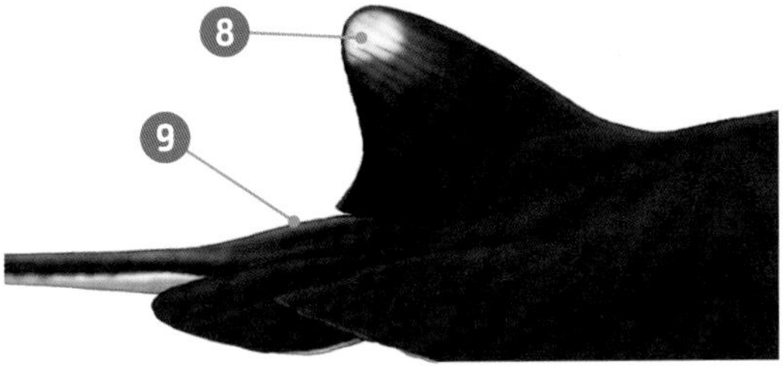

8 9

Lateral view of tail region

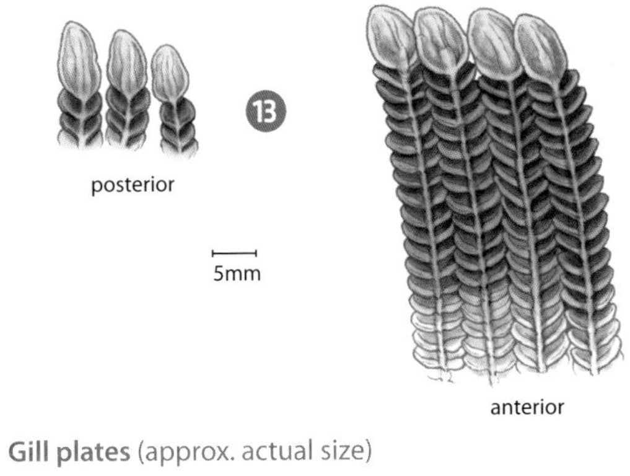

lateral

posterior

13

5mm

anterior

Gill plates (approx. actual size)

11 12

Lateral view of head

Caribbean Manta Ray *Mobula* cf. *birostris* (not illustrated)

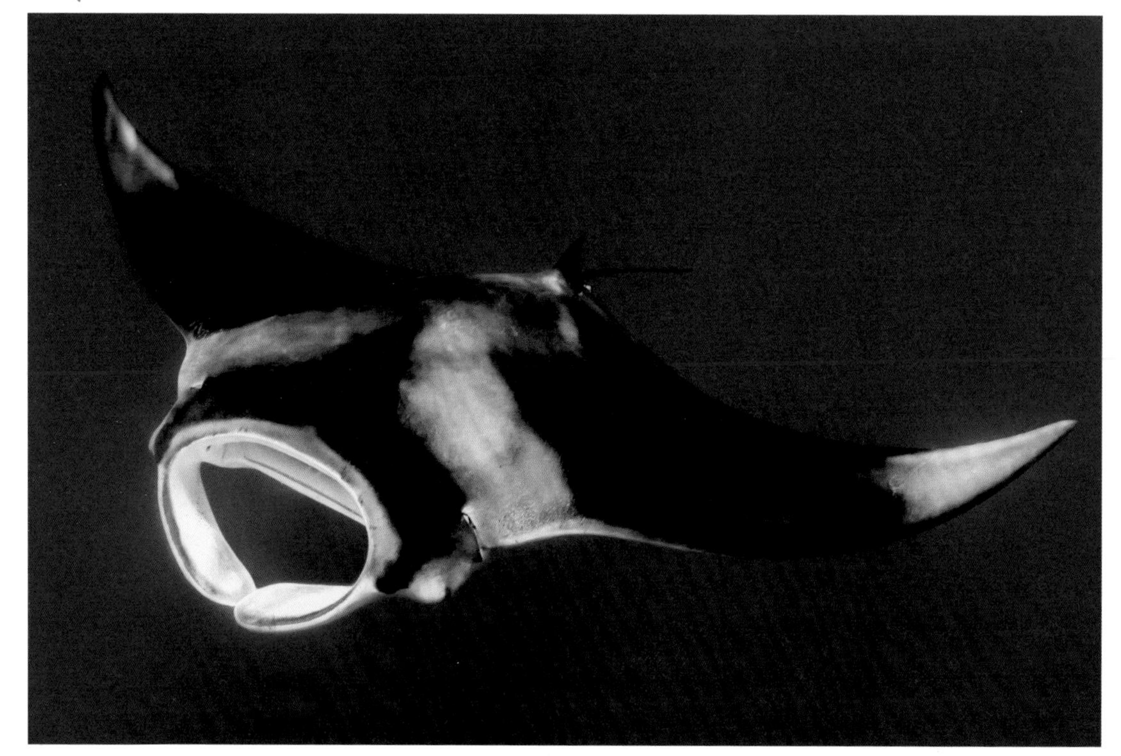

Caribbean Manta Ray, *Mobula* cf. *birostris*, Isla Mujeres, Mexico.

Species characteristics

Disc width: maximum 500cm (16.5ft), average 300–400cm (10–13ft).

Weight: up to 1,000kg (2,200lbs).

Size at maturity: unknown, but likely to be smaller than Oceanic Manta Rays.

Age at maturity: unknown, but likely to be similar to Oceanic and Reef Manta Rays.

Lifespan: likely to be around 40 years.

Reproduction: unknown, but likely to be similar to Oceanic and Reef Manta Rays.

Distribution: found throughout the Caribbean Sea and the Gulf of Mexico; its range possibly extending further throughout the Atlantic Ocean.

IUCN Red List: **Vulnerable**

FAO species code: **RMB**

Throughout the reef habitats of the Caribbean Sea and Gulf of Mexico there occurs a proposed third species of manta ray that is sympatric to the Oceanic Manta Ray in this region. This putative 'Caribbean' Manta Ray *Mobula* cf. *birostris* appears to occupy a similar niche to the Reef Manta Ray. It may well therefore be that some of the pioneering Oceanic Manta Rays that crossed the open ocean to the Caribbean found plentiful sources of food along the inshore reefs and began to diverge into a new species – essentially, nature replicating the evolution of Reef Manta Ray again. However, although recent genetic analysis supports a degree of separation between these two groups, there remains uncertainty on the validity of this proposed speciation.

The putative Caribbean Manta Ray is smaller than its oceanic cousin and more similar in size to the Reef Manta Ray. The coloration dorsally is a cross between Reef and Oceanic Mantas. Unlike in Oceanic Manta Rays, where the black 'T'-shaped head marking extends down through the back to join the rest of the black-coloured body, maintaining a roughly even width throughout, in the Caribbean Manta the lower tail of the 'T' tapers together towards the bottom where it joins the back.

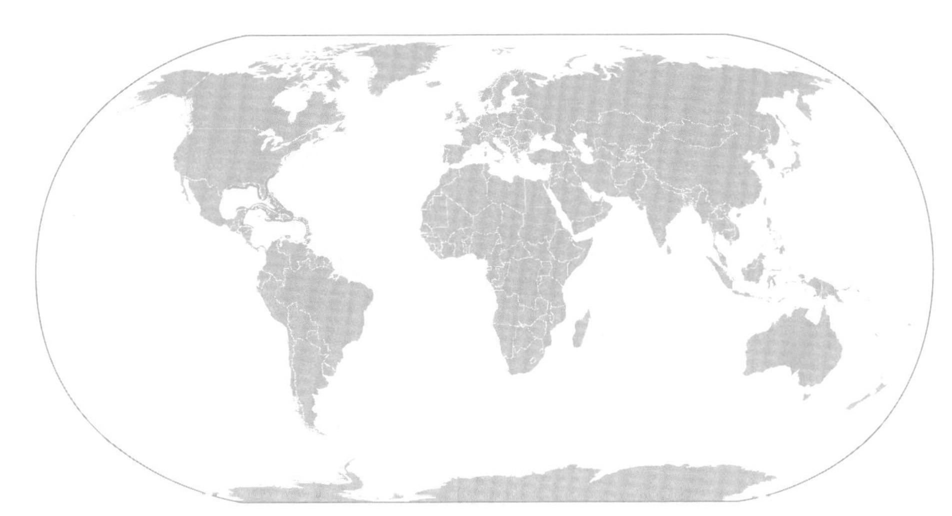

Distribution map of the Caribbean Manta Ray *Mobula* cf. *birostris*. Darker areas indicate confirmed range; lighter areas indicate expected range.

In Reef Manta Rays this tapering of the 'T' tail is even much more acute, often forming instead more of a 'V' shape. Ventrally, the putative species has coloration which is more similar to the Oceanic Mantas, lacking for the most part black spots between the gills and across the pectoral fins. Instead, if any spots are present (as in oceanics), they are usually clustered around the belly below the gills. Caribbean Mantas also often have large patches of round or oblong shaded areas on the belly below the gills. These grey-shaded areas often also extend right across the pectoral fins ventrally and up around the head, mouth and gills, increasing coverage with age.

A juvenile Caribbean Manta Ray *Mobula* cf. *birostris* swims along Mount Irvine Extension Reef off Pigeon Point in Tobago. This 'Caribbean' manta has been proposed by some scientists as a third manta species. However, while there are some habitual and coloration differences, there remains some uncertainty on the validity of this proposal-based genetic analysis.

MOBULIDS AND PEOPLE

For many tourists, snorkelling or diving with a manta ray is an experience that connects them to the ocean and its inhabitants for the first time.

Threats to mobulids

The greatest threat to mobulids is excessive targeted and incidental take in fisheries, increasingly driven by the international trade in gill plates for use in an Asian health tonic purported to treat a wide variety of conditions. As a result, some mobulid populations have exhibited regional declines of over 80%. Of particular concern is the exploitation of this species from within areas of critical habitats, where numerous individuals can be targeted with relatively high catch-per-unit-effort.

Entanglement in mooring lines, marine debris and boat strikes also injure mobulid rays, decrease fitness or contribute to non-natural mortality. Additional threats include habitat destruction, pollution and climate change. For such intrinsically vulnerable species, even small negative pressures exerted upon a population are likely to have severe consequences for the population's survival.

Indirect threats to manta and devil rays kill thousands of these animals every year. These rays are easily entangled in fishing nets and lines (bottom right), or mooring ropes (middle), or hit by speed boats – the outboard engines' propellers slicing through the animal's body (top and bottom left). Ironically, these indirect threats often increase in areas where manta and devil ray tourism activities develop, as increased boat traffic, mooring lines and fishing activities are required to satisfy the demands of the visiting tourists.

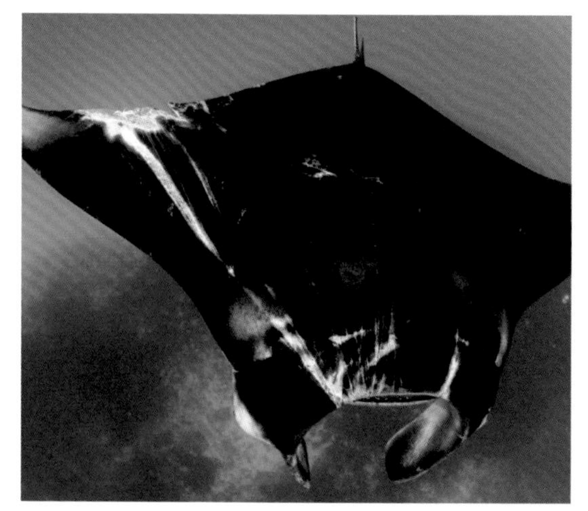

Fisheries

One of the first countries to commercially fish their mobulid ray populations was Mexico. In the 1980s fishermen in the Sea of Cortez switched from subsistence bycatch fishing of the locally abundant Oceanic Manta and devil ray species to directed target fisheries. They used harpoons to impale the surface-feeding animals, and gill nets to entangle them. The rays were easy targets and their numbers soon began to plummet. The giant carcasses were towed back to the beaches where the choicest flesh was sold for consumption, while the remainder was often used as bait in lobster pots or simply discarded.

Within just a decade the manta ray population in the Sea of Cortez was virtually wiped out and the fishery collapsed. It was not until 2006 that the Mexican government finally passed legislation protecting Oceanic Manta Rays in Mexican waters, but by then the damage had already been done. Even today, after many years of protection, very few mantas are recorded in this area, and those that are, still fall victim to illegal fishing or bycatch.

Many other countries have since targeted their manta and devil ray populations, switching from small-scale local consumption fisheries to commercial fisheries wherever a market for their products can be found. The Philippines, Indonesia, Mozambique, India, Sri Lanka and China have all targeted their mobulid ray populations in recent decades. Where these targeted fisheries have occurred, similar trends of population declines to those reported in Mexico have also been observed. Yet of all these countries, only the Philippines and Indonesia have official laws in place, often with limited implementation, to protect these vulnerable species.

The reason for these rapid declines is simple: manta and devil rays live for a long time and reproduce infrequently. They are large animals with few natural predators and have long gestation periods that result in the birth of just a single pup (most of the time). The pups themselves take a long time to reach sexual maturity.

As a result of these life history strategies, and like most other large marine animals, manta ray populations

Targeted fisheries for manta and devil rays occur throughout their range, killing tens of thousands of individuals annually for their gill plates and meat, which are cut into strips and consumed fresh (second from bottom) or salted and sun dried (bottom).

(and probably most of the devil ray species also) simply cannot survive, or sustain, any commercial fisheries for long. Any target fishery that annually removes even a small percentage of the breeding adults results in a rapid decline in the overall population within just a few years because the remaining mature individuals cannot breed fast enough to replace the losses.

This is why, even with complete protection from anthropogenic threats, any overfished population of these rays will take many decades to recover to its natural state. In the realities of today's global fisheries management and protective enforcement (or lack thereof), this is unlikely ever to happen where populations have already been overfished.

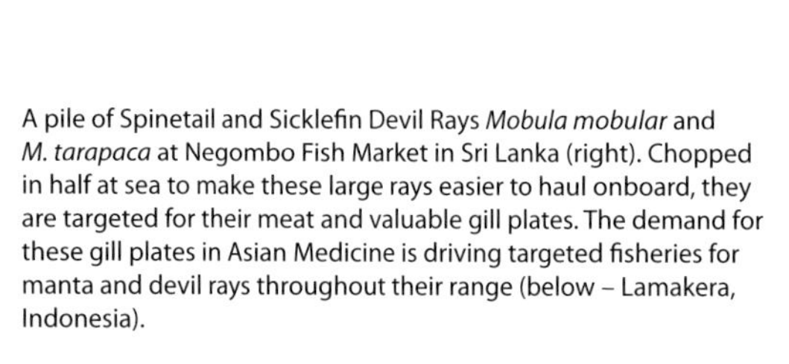

A pile of Spinetail and Sicklefin Devil Rays *Mobula mobular* and *M. tarapaca* at Negombo Fish Market in Sri Lanka (right). Chopped in half at sea to make these large rays easier to haul onboard, they are targeted for their meat and valuable gill plates. The demand for these gill plates in Asian Medicine is driving targeted fisheries for manta and devil rays throughout their range (below – Lamakera, Indonesia).

Gill plate trade

While the meat from the fished rays is often consumed locally in either fresh or dried form, the driving factor behind the dramatic increases in fishing efforts for manta and devil rays, particularly throughout South East Asia in recent decades, is the growing demand for their gill plates. These are dried and exported for the Asian medicinal market. Small pieces of the dried plates are added to a broth, along with crushed pipefishes and ginseng, to be used as a treatment for a variety of ailments. The gill plates are marketed as having anti-inflammatory properties, clearing away heat and toxic material, and eliminating stasis to activate blood circulation.

There is absolutely no scientific evidence to back up these medicinal claims and the first reference in the Traditional Chinese Medicine texts that list manta or devil rays' gill plates as being used for medicinal purposes dates back only to 1976. This pseudo medicine, it seems, is fairly new – clever marketing of a readily available and cheap bycatch product. This so-called 'medicine' targets those willing to believe that because the ray's gill plates can filter the water for food, when consumed by humans, the gills will filter and remove toxins from their bodies. The recent increase in demand for mobulid gill plates may also be linked to the increase in air pollution in China over the last few decades. With more people now suffering from respiratory and other associated air pollutant illnesses, this new 'medicine' takes advantage of Asian people who have a deep-rooted belief in the healing properties of traditional medicines. Furthermore, the growing Chinese economy means that many more people are able to afford this product.

Regardless of the precise reasoning and driving force behind this trade, it is clear that many people in Asia are still prepared to pay large sums of money to buy these gill plates and as long as there is a demand, there will always be an incentive for fishermen to catch and kill the rays to supply it. Unfortunately, these senseless boom-and-bust fisheries serve only to line the pockets of a few individuals, exporting the natural wealth of developing nations for short-term rewards. Such fisheries impoverish the masses and squander the nations' natural heritage.

Manta ray gill plates laid out to dry in the sun.

We are rapidly depleting our oceans of these fascinating animals for nothing more than a gimmick product, the result of entrepreneurial marketing that exploits cultural beliefs. Furthermore, recent studies have revealed the presence of unsafe levels of heavy metals within many gill plates tested, exceeding China's pharmacopoeia recommended limit for arsenic by as much as 20 times.

The gill plates are cut from the head of the ray (top left), then laid out in the sun to dry for several days (middle and bottom left) before being shipped for sale in Asian retail shops, mostly in China, marketed as an ingredient in traditional medicines (below).

Bycatch and discard

The term bycatch is used to describe fish or other animals that are caught unintentionally in a fishery. In 1997, the Organisation for Economic Co-operation and Development defined bycatch as 'total fishing mortality, excluding that accounted [for] directly by the retained catch of target species'.

While manta and devil rays are increasingly being fished and targeted for their gill plates in some countries around the world, bycatch of these large pelagic rays in high seas fisheries is still seen as worthless and the bycaught rays are discarded. Each year thousands of manta rays and many tens of thousands of devil rays are caught as bycatch in the global hunt for more desirable species. The captured animals are often already dead, dying or seriously injured before being thrown back into the sea.

Each year the world's fisheries discard seven million tonnes of fish – that is one-tenth of the world's global catch thrown back over the side of the boat. Some of the most destructive forms of fishing, such as drift nets (now illegal in international waters and within many nations' territorial waters), longlining and gill-netting, are responsible for much of this waste. They indiscriminately catch and kill huge numbers of some of our oceans' most endangered species, from dolphins and whales to sea turtles, albatrosses and sharks, as well as the manta and devil rays. The Oceanic Manta Ray and all of the larger devil ray species are especially vulnerable to these oceanic fisheries, falling victim to illegal drift nets, gill nets and tuna purse seine fisheries in vast numbers each year as they migrate across the tropical and subtropical oceans of the world.

A juvenile Reef Manta Ray *Mobula alfredi* has become entangled in a gill net in the Maldives. With its left cephalic fin severely damaged, this little manta's chances of survival are poor. All over the world, fishing nets and lines are killing countless marine creatures which become the unintended discard and bycatch victims of our growing human demands on the ocean's resources.

Conserving mobulids

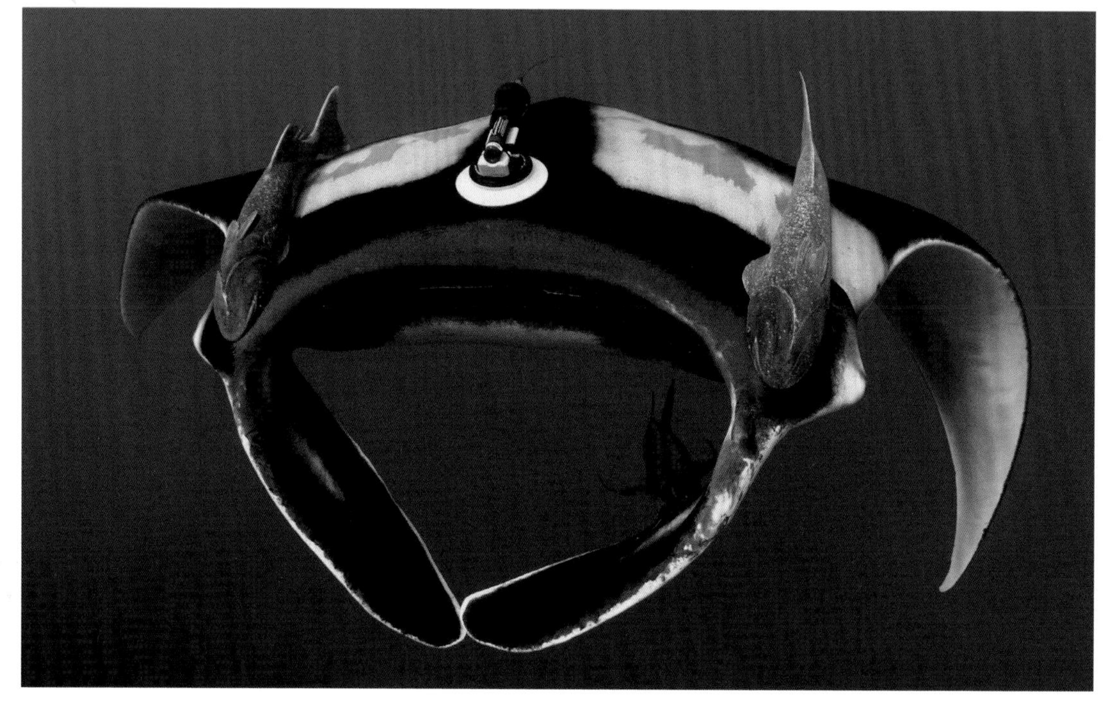

Like a man-made remora, a Crittercam sucks on to the head of an Oceanic Manta Ray *Mobula birostris* in Revillagigedo Archipelago Marine Protected Area. While attached, the Crittercams record everything the manta sees, where it goes and what it does.

Most of Earth's surface is covered not by land but water, the planet's oceans spanning almost three-quarters of our home. Free for all to use and benefit from, these vast areas of ocean are the birth place of life on Earth and we depend on them for our survival in more ways than most of us imagine. Unfortunately, most nations and individuals still view our oceans and their inhabitants as nothing more than a common resource; a dark reservoir of infinite depth to exploit and plunder, each selfishly taking more than their fair share, raping the natural world of more than it can provide in a bid to extract national and personal profits from the ever-diminishing reserves of our world bank.

After thousands of years of exploitation, our world is full of unnaturally empty seas, impoverished and stripped of their natural resources as one species after another has been fished to the point of commercial extinction for human greed. This dilemma is termed a 'Tragedy of the Commons', a notion first conceived and published by Garrett Hardin in the journal Science in 1968. His article describes a situation in which multiple individuals, acting independently and rationally consulting their own self-interest, will ultimately deplete a shared limited resource even when it is clear that it is not in anyone's long-term interest to do so. Unless the world can unify and agree to work towards the common and mutually beneficial goal of conserving our oceans' natural resources, then the planet we leave to our children will be far less rich than the one we inherited.

Confounding this tragedy is the general lack of awareness and care surrounding these issues by the general public. Humans are land lovers; for most of us the oceans are alien and uninviting places. What goes on under the seas is disconnected from our everyday lives, easily forgotten or brushed aside, out of sight and out of mind. Aside from selfish personal gains, humans will generally protect only what we care about and can connect with. 'In the end we will conserve only what we love. We will love only what we understand [and] we will understand only what we are taught' (Baba Dioum). So to truly conserve a species or its habitat, there must first be the desire to do so, the passion driven by empathy which arises from care, understanding, and above all, knowledge.

Science is the tool that enables conservationists and governments to make informed and worthwhile decisions that can help to curb the growing pressures exerted on our planet's oceans and their inhabitants. To preserve a species or its ecosystem you must first understand it – how it functions, what areas and aspects of its life are critical to its survival, and what impact humans have on these factors.

Currently, very little is known about the life history strategies of mobulid rays, especially the devil rays. Fundamental questions such as how long they live, their age at maturity, their reproductive cycles and fecundity, have yet to be answered satisfactorily. Areas of key habitat use, migration corridors and population estimates, both nationally and internationally, must also be clearly defined if we are to make the informed and educated decisions needed to effectively protect these animals and the world they inhabit.

For many people the mere mention of the word 'shark' or 'devilfish' conjures up images of ferocious killers roaming our oceans' depths ready to savage any human who dares to enter their realm. These preconceived notions are far from reality, born of ignorance, fear of the unknown and an active imagination which stirs within us all. The truth is that sharks and rays are not demons with sinister desires or evil motivations.

The more we delve into their world and begin to unravel the mysteries of their lives, the more we realise just how misinformed our preconceptions of these animals are. Manta and devil rays, whale sharks and the vast majority of all other elasmobranchs are harmless to humans. As this realisation spreads and human perceptions begin to change, millions of people around the world each year collectively spend billions in search of their own close encounters with animals that just a few decades ago were generally perceived as dangerous sea monsters.

Tourism has opened up many new opportunities to exploit and profit from our oceans, which in turn can and should be used to protect them. And while tourism is certainly less environmentally destructive to our oceans than fishing, it nonetheless creates its own problems that need to be tackled. As more and

Raising awareness and making people care about our oceans and their inhabitants is key to successful conservation efforts. A delegate watches a virtual reality film about devil rays at the CITES conference of parties in South Africa, were these species were proposed for protective measures (top). Manta Trust staff educate tourists at a marine festival in the Maldives (middle), while children in the Philippines learn about manta and devil ray species identification and conservation (bottom).

more people come into contact with mobulid rays and all the other creatures of our oceans the tide of ignorance is slowly starting to turn towards that of awareness, leading to a wider call for global protection and conservation of our planet's oceans and their inhabitants.

Today 40% of the world's human population live within just 100 kilometres (62 miles) of the shore. Millions of people depend directly on the oceans and their natural resources for their income and daily survival. It is these people with the most to lose who will inevitably lead the way in efforts to highlight the most pressing issues facing our oceans globally, from overfishing to sea level rise, ocean acidification, pollution and coral bleaching. It is their voice that will shout the loudest and hopefully pave the way for a more sustainable and holistic approach to the management and preservation of our global marine resources which, until now, have been taken for granted without regard for the consequences of their loss.

It is in these coastal communities that mobulid rays are now an important part of a regional and national tourism industry, driving multi-million dollar revenue streams. The night-feeding manta rays at Kona on Hawaii's Big Island generate over US$11 million annually for the local economy, and in the tiny island nation of the Republic of Maldives the reef manta population generates a massive US$15 million annually for the local economy. Globally it has been estimated that direct revenue generated from manta ray tourism is US$73 million annually and direct economic impact, including associated tourism expenditures, is close to US$140 million annually. Just ten countries account for 93% of the manta-related revenue generated: Japan, Indonesia, Maldives, Mozambique, Thailand, Australia, Mexico, United States, Palau and Federated States of Micronesia.

This huge global revenue stream from manta-related tourism means that every individual manta ray is worth many thousands of dollars alive. They are veritable marine cash cows grazing the oceans' plankton and generating vast sums of renewable cash for the local economies of these nations. Furthermore, manta rays live for many decades so the total revenue each manta ray can generate during its lifespan is immense. By comparison, a single dead manta ray's meat and gills will sell, at most, for just a few hundred US dollars for the fishermen who caught it – a one-time deal, the true costs of which are incalculable, both in terms of future revenue lost and ecological impoverishment.

Researchers tow a plankton sampling net to identify and quantify the zooplankton community in the water column (top). A researcher uses a stereo video system to measure the size of a manta in the Maldives (bottom).

Although tourism can provide a much better socio-economic alternative to intensive fishing practices, it still brings with it a wide variety of problems that need to be managed. As tourism continues to grow in nations where manta rays occur, the impacts on these tropical destinations and their coral reefs are also increasing. To reduce these negative impacts it is essential for the whole community to be engaged so they are made aware of the problems and become a part of the solution. In a world increasingly driven by commercialism, the connection has to be made between the need to protect these charismatic megafauna and their habitats, and the economic revenue that will ultimately flow into each person's pockets from their continued preservation.

To achieve real conservation of a species or its habitat there must first be the will to do so by those people who directly create the impact. Regulations and laws that underpin society will always be broken by those few selfish individuals who do not accept or respect the wider long-term benefits to all, while apathy and ignorance will also lead to inertia and failure. Therefore, widespread education and awareness about the importance of protecting our oceans is integral to its success. Tourists, business stakeholders and, most importantly, the local communities, must understand and believe in these initiatives themselves.

Local communities need to feel a sense of pride and passion for their natural heritage. Historically, coastal communities have always been intrinsically linked to, and dependent on, their surrounding seas. But as the influences of the outside world become stronger with each growing generation, the connection with our oceans has been slowly eroded, undermining the foundations of a sustainable approach to utilising our marine resources and threatening these communities every bit as much as climate change, pollution and overfishing.

Encouragingly, community and government-led educational initiatives are springing up all over the world, and they are a vital step along the road that leads back to awareness, and in turn real conservation. We need to lay the foundations for a future generation that will be more in tune with nature and the impacts we all have on our natural world – a world we have taken so much for granted. There is still time to save it if we care enough to make it happen.

Local children in Peru pose for the camera in front of a manta ray mural they helped to paint as part of a community awareness programme.

Protective legislation

While international, regional and national protective legislation has improved in recent years, there is still a greater need for protection throughout the range of all manta and devil ray species globally. A few countries have taken the lead in mobulid ray conservation, protecting these species to differing degrees within their territorial waters.

Location	Species
International	
Convention on International Trade in Endangered Species (CITES)	All mobulid species
Convention on the Conservation of Migratory Species of Wild Animals (CMS)	All mobulid species
Inter-American Tropical Tuna Commission (IATTC)	All mobulid species
Regional	
Barcelona & Bern Conventions	*M. mobular*
European Union member countries	All mobulid species
National	
Australia	All mobulid species
Brazil	All mobulid species
Croatia	*M. mobular*
Ecuador	*M. birostris, M. mobular, M. thurstoni, M. munkiana* and *M. tarapacana*
Indonesia	*M. birostris* and *M. alfredi*
Israel	All ray species
Maldives	All ray species
Malta	*M. mobular*
Mexico	*M. birostris, M. mobular, M. thurstoni, M. munkiana, M. hypostoma* and *M.*
New Zealand	*M. birostris* and *M. mobular*
Peru	*M. birostris*
Philippines	*M. birostris* and *M. alfredi*
United Arab Emirates (UAE)	*M. birostris* and *M. alfredi*
State	
Christmas Island and Cocos (Keeling) Islands, Australian Indian Ocean Territories	All ray species
Commonwealth of the Northern Mariana Islands, USA Overseas Territory	All ray species
Florida, USA	*M. birostris* and *M. alfredi*
Guam, USA Overseas Territory	All ray species
Hawaii, USA	*M. birostris* and *M. alfredi*
West Manggarai/Komodo, Indonesia	*M. birostris* and *M. alfredi*
Raja Ampat Regency, Indonesia	All ray species
Yap, Federated States of Micronesia	*M. birostris* and *M. alfredi*

Not surprisingly, the most conservation minded of these nations are also those that derive the most economic benefits from manta and devil ray tourism.

Legal protection measure
Listing of the genus *Manta* (2013) and *Mobula* (2016) on Appendix II.
Appendix I and II; *M. birostris* (2011), all other mobulid species (2014), Sharks MoU (2016) and Concerted Actions for Mobulid Rays (2017).
Resolution C-15-04 on the Conservation of Mobulid Rays Caught in Association with Fisheries in the IATTC Convention Area.
Added to the Annex II 'list of strictly protected fauna species' of the Bern Convention and the Annex II 'List of endangered or threatened species' to the Protocol concerning Special Protected Areas and Biological Diversity in the Mediterranean of the Barcelona Convention, which came into force in 2001.
Council Regulation (EU) 2015/2014 amending Regulation (EU) No 43/2014 and repealing Regulation (EU) No 779/2014.
Environment Protection and Biodiversity Conservation Act (added as protected species 2012).
Inter-ministerial Normative Instruction No. 2 of 14/3/2013.
Law of the Wild Taxa 2006 Strictly prohibited.
Ecuador Official Policy 093, 2010.
KepMen National Protective Legislation, 2014.
All sharks and rays fully protected in Israel since 2005. They may not be captured, harmed, traded or kept, without a specific permit from the Israel Nature and Parks Authority (INPA).
Exports of all ray products banned 1995. Environment Protection Agency rule – illegal to capture, keep or harm any type of ray; Batoidea Maldives Protection Gazette No. (IUL) 438-ECAS/438/2014/81.
Sch. VI Absolute protection.
NOM-029-PESC-2006 Prohibits harvest and sale.
Wildlife Act 1953 Schedule 7A (absolute protection).
Article 2 of Resolution 441-2015-PRODUCE, Jan 2016.
FAO 193 1998 Whale Shark and Manta Ray ban.
Fully protected in UAE waters (2014).
Protected species. Dept. of Fisheries Western Australia 2010.
Public Law No. 15-124.
FL Admin Code 68B-44.008 – no harvest.
Bill 44-31 prohibiting possession/sale/trade in ray parts 2011.
H.B. 366 2009 – no harvest or trade.
Shark and Manta Ray Sanctuary Bupati Decree 2013.
PERDA (Provincial Law) Hiu No. 9 Raja Ampat 2012.
Manta Ray Sanctuary and Protection Act 2008.

Marine Protected Areas

The Oceanic Manta Rays *Mobula birostris* at El Boiler, a dive site within Mexico's Revillagigedo Archipelago MPA, are some of the most engaging marine animals anywhere in the world.

A species' habitat is its home; it depends on it for shelter and searches within it for food and the opportunity to mate. To effectively ensure the long-term survival of a species in the wild you must also ensure the protection of its habitat, especially those areas of critical importance to its survival, such as mating, feeding or birthing grounds. Protecting habitats to safeguard a specific species also helps to ensure the protection of all the other species within that habitat that also depend on it for survival. However, life on Earth is usually not so easily and discretely defined or protected, with varying scales of interconnectivity for each species occurring between habitats over space and time. This is especially true for large migratory species like mobulid rays, which often range across both natural and human borders, making complete protection for these species extremely difficult without in-depth knowledge of their lives.

Identifying the areas of habitat that are most important for the protection of a species is just the first step on the road to successful and effective protection. The next step – designating these areas for protection – is an even more difficult process, often met with conflict and political frustrations. Once protective legislation exists on paper, the task of enforcing and effecting the sustainable management of these regulations is essential to gain real conservation benefits. Around the world there are several Marine Protected Areas (MPAs) that have been created principally to ensure the protection of the local population of manta rays, while others, created less specifically for the rays, equally serve as important refuges for these animals.

The IUCN defines an MPA as 'any area of the intertidal or subtidal terrain, together with its overlying water and associated flora, fauna, historical and cultural features, which has been reserved by law or other effective means to protect part of, or the entire, enclosed environment'.

Unfortunately, global marine resources are still viewed almost entirely as open markets for anyone to utilise and benefit from; in 2017 only 13% of the world's oceans were protected, and most of these protected areas still allow fishing to some degree within their boundaries. Top fisheries management scientists suggest that if we are to truly preserve our oceans and ensure sustainable fisheries for the future population of the Earth, one of the key changes to the way we manage our oceans must be the creation of many more protected areas. They recommend that approximately 30% of our oceans' key marine habitats must be set aside from fishing and other destructive pressures, with effective enforcement of the protective legislation also essential for success.

Of course, recommendations like these are usually met by opposition from fishing communities and international commercial fishing enterprises, which claim this will reduce the amount of fish they will be able to catch. However, research has repeatedly shown that the opposite often occurs, with spill-over catches occurring in the surrounding waters outside the MPA increasing the total catches for the fishermen, while still preserving the core protected areas and species populations within the MPA.

And the simple truth is that we have to change the way we fish our oceans. If we carry on as we are, there will be no fish left for the fishermen to catch anyway; we are rapidly approaching the end of the line and the only catch will be the final realisation of our own ignorance and selfish greed.

As nations' natural resources continue to shrink while their economies and populations grow, MPAs are one of the best ways of helping to alleviate the various pressures exerted on the surrounding ecosystems and their inhabitants as a result of these changes. However, protection is easy to proclaim but less easy to achieve and sustain effectively. Too many of the world's current MPAs are in reality nothing more than paper parks – borders that have been easily outlined on maps, the area's contents declared protected by well-meaning governments, but that in reality are often no more protected than any other area of the oceans.

Manta and devil ray sanctuaries

Listed below are a few examples of important MPAs which, through their creation, have helped to safeguard areas of critical devil ray and manta habitat.

Mexico The Revillagigedo Archipelago is a group of four volcanic islands in the Pacific Ocean known for their unique ecosystem. These remote islands lie over 400 kilometres (250 miles) southwest of the southern tip of Baja California in the Pacific Ocean and were designated a marine reserve by the Mexican government in 2017, the largest protected area of its kind in North America. The no-take area of the reserve encompasses 148,000 square kilometres (57,000 square miles). The population of Oceanic Manta Rays at this remote location are still relatively healthy when compared to other areas along the mainland coast of Mexico where intensive fishing pressures in the past dramatically reduced the manta population.

Indonesia The Komodo National Park is a good example of a protected area that has afforded manta and devil rays protection as a result of its existence. Now a UNESCO World Heritage Site and Biosphere Reserve, the park was set up initially as a terrestrial park in 1980. Since its inception, its remit has expanded to protect not only terrestrial biodiversity but also marine. A total of 67% of the protected area is now marine, giving refuge not only to manta and devil rays but all other animals that inhabit this rich coral reef ecosystem. In 2012 the Raja Ampat Regency created a 44,000 square kilometre (17,000 square mile) sanctuary in the western province of Indonesia, protecting all sharks, mobulas, turtles and dugongs in the area. As marine-related tourism continues to grow in Indonesia it is hoped that many more large MPAs such as these will be created to ensure the protection of these habitats through the economic incentives brought about by tourism.

A school of Munk's Pygmy Devil Rays *Mobula munkiana*, numbering several hundred individuals, observed off the coast of the Baja Peninsula in Mexico, where these rays are now protected.

Yap In the small island state of Yap in Micronesia the resident population of Reef Manta Rays is very important for the local tourism industry. As a result, all manta and devil ray species are heavily protected, with anyone found harming these animals facing a six-month prison sentence and a US$1,000 fine. A manta ray sanctuary has also been created in Yap; covering 21,325 square kilometres (8,234 square miles) of ocean it encompasses 16 islands and 145 islets, extending 20 kilometres (12 miles) offshore.

Maldives The Republic of Maldives boasts a total of 34 marine and terrestrial protected areas. The three most recently designated of these MPAs (announced in June 2009) were specifically chosen because of their importance as key sites of habitat use for manta rays and whale sharks. Unfortunately, before these 2009 designations regulatory measures set in place on paper to protect these areas were poorly enforced. However, the MPAs designated in 2009 are heralding a new era for the management of MPAs in the Maldives, with site-specific management plans and on-site enforcement now in place at several sites that will hopefully be expanded to the remaining areas in the future. The most well-known of these MPAs is Hanifaru Bay in Baa Atoll. This is a key feeding area for Reef Mantas during the Maldives' southwest monsoon and despite its small size (around the size of a football pitch) the bay has gained worldwide recognition for its ability to attract as many as 250 mantas into its sandy shallows during peak feeding events.

Chagos In 2010 the British Indian Ocean Territory was declared an MPA by the British government, intended to be the largest contiguous no-take marine reserve in the world, spanning an area of 640,000 square kilometres (397,667 square miles). Located 500 kilometres (311 miles) south of the Maldives in the Indian Ocean, the Chagos Archipelago has been uninhabited since 1970 (except for military personnel on Diego Garcia Atoll). As a result of its remoteness and the lack of human activity throughout the archipelago, the Chagos marine reserve contains one of the healthiest reef systems and least polluted waters in the world. The reserve is home to a healthy population of Reef Manta Rays; through the protection of their habitat they inhabit a safe haven in the centre of the Indian Ocean.

St Helena and Ascension Islands In 2016 a 200 nautical mile maritime zone was created around the remote island of St Helena, a British Overseas Territory (BOT) in the Atlantic Ocean. The UK government has also pledged that at least half of the 445,390 square kilometres (171,966 square miles) marine environment around Ascension Island (another BOT, 700 nautical miles north of St Helena) will be permanently closed to commercial fishing by 2019. St. Helena's and Ascension Island's maritime environments are home to over 40 endemic species and supports a diverse array of marine life including Whale Sharks, Humpback Whales and sea turtles. The waters around these islands also appear to be important aggregation sites for Sicklefin Devil Rays *Mobula tarapacana*, which gather at remote islands and seamounts in the Atlantic to engage in reproductive and feeding activities. These oceanic rays are particularly vulnerable to high seas fisheries, such a tuna purse-seines, therefore these protected areas offer an important reserve for these highly migratory species, which have been tracked using satellite tags travelling thousands of miles between oceanic islands in the Atlantic.

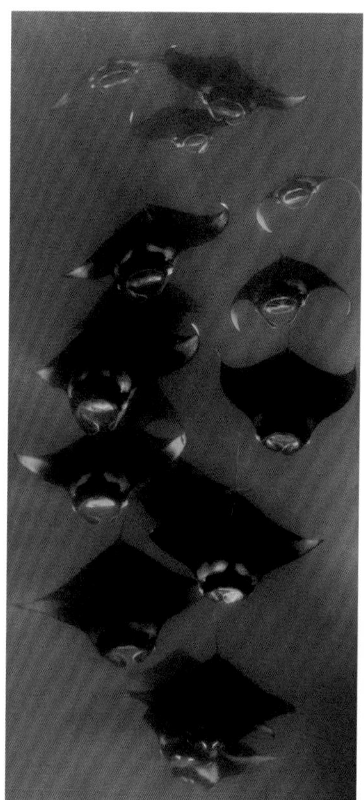

Manta ray cleaning stations, like this one off the 'manta sanctuary' island of Yap in Micronesia, serve as important focal points for courtship and mating, as well as cleaning (top). Spectacular Reef Manta Ray *Mobula alfredi* chain-feeding events like this are a common occurrence inside Hanifaru Bay MPA in the Maldives (bottom).

Citizen science

Over the last few years significant advances have been made in understanding the lives of manta and devil ray populations all around the world. Indeed, this guide is a testament to the knowledge that has been gained by researchers and organisations working in those areas. However, there is much still to be learned, and everyone – not just scientists – can contribute to the science by collecting photo identification (photo-ID) images of manta and devil rays when they are seen. To date, tourists and researchers have contributed tens of thousands of images to the Manta Trust's global database, enabling their researchers to estimate population sizes, plot migration routes and species' ranges, map reproductive patterns and determine the demographics of the populations.

For the manta species, photo-ID techniques are especially useful because every individual has its own uniquely identifiable pattern of black spots and blotches on its belly (ventral surface). Just like a fingerprint, this does not change throughout the animal's life. Using this research technique in the Maldives alone, researchers have built the world's largest manta ray photo-ID database, which now contains over 50,000 sightings of more than 4,500 individual manta rays collected from across the country's 26 geographical atolls. This dataset is a significant tool that has been used to help draw important conclusions about the Maldives' manta population, which in turn have be used to make informed conservation recommendations.

While the photo-ID work is an integral part of the researchers' work, there are other methods of data collection that are just as helpful and important in understanding these animals. Satellite, acoustic and other forms of tagging are vital to help define both large-scale migration patterns and small-scale daily movements, tracking the animals as they swim through the water column, both vertically and horizontally, and allowing scientists to define areas and habitat usage.

Genetics, stable isotope studies, behavioural observations, life history and anthropogenic (human) impacts studies are also important aspects of the global research that helps scientists to build a complete picture of the manta and devil rays' lives in order to better protect and conserve them for the future.

Divers and snorkellers flock to manta and devil ray aggregation sites around the world to enjoy the thrill of sharing the ocean with these graceful giants, but it is essential for this tourism to be managed so as to minimise the negative impacts on the rays and their habitat.

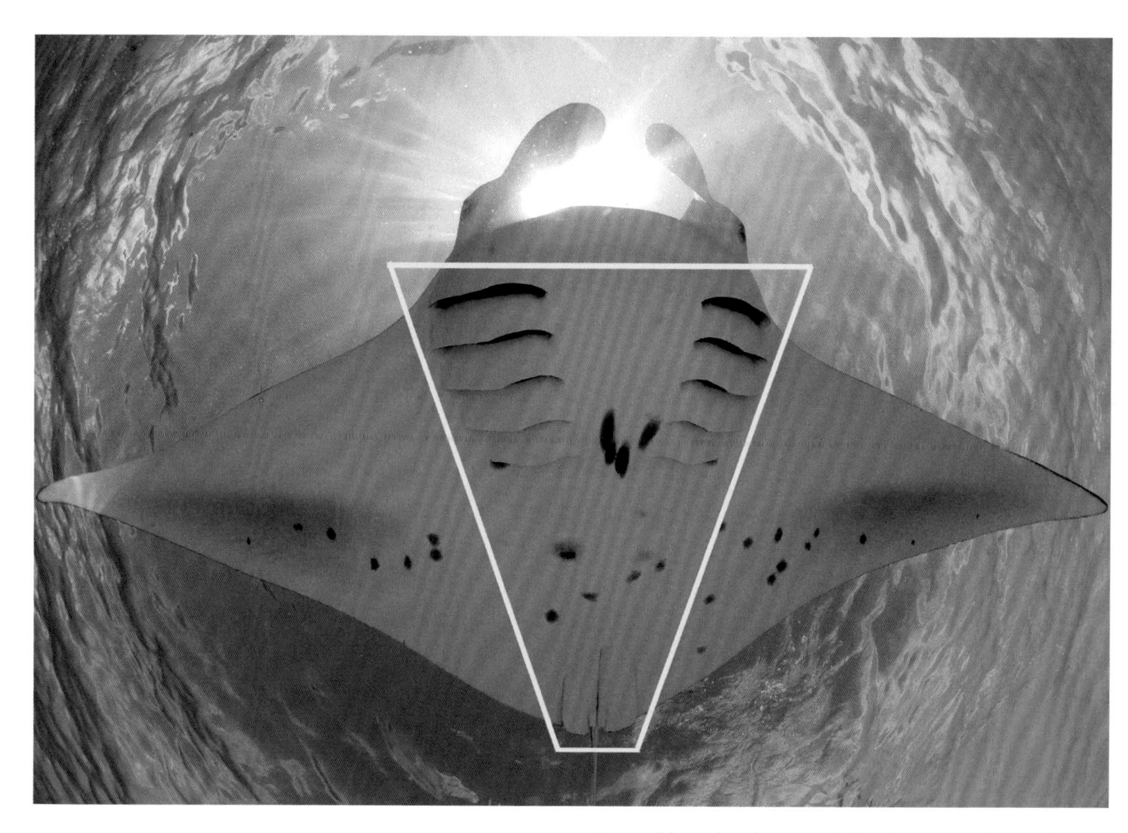

It is possible to identify every individual because each manta has its own unique pattern of black spots on its predominantly white belly. These patterns do not change throughout the lives of the mantas, enabling researchers to track each individual as it is sighted over the decades (mantas live for more than 40 years). Every manta sighting – whether of a new manta or a re-sighting of an individual that is already known – is an important piece of a huge jigsaw puzzle.

IDtheManta – Global Database

IDtheManta is a global manta ray photo-ID database platform in which automated visual biometric photo-ID technology is linked to a global relational database accessible to both manta scientists and the general public. It is a massive data source for scientists, enabling research organisations to monitor and learn about the manta ray populations around the world. *IDtheManta* also helps to raise awareness and drive the conservation of manta rays and their habitats globally by providing in-depth feedback to every individual who uploads a sighting to the database.

Anyone can contribute directly to the science and conservation of manta rays by submitting their images directly through the Manta Trust's website **www.mantatrust.org**. Ideally, we are looking for images that best show the spots on the underside (ventral surface) of the manta rays. Other images that show the top (dorsal surface) of the manta rays, or the tail area, can also be used to identify the specific species you encountered and/or the sex of the individual.

Swimming with mobulids

There are few experiences more exciting than diving or snorkelling with manta and devil rays. Every year tourists spend an estimated US$140 million globally to see manta rays in the wild. Tourism is potentially one part of the solution to the issue of global manta fisheries, providing many countries with a strong economic incentive to protect these animals and fishing communities with an alternative livelihood. By swimming with mobulids you could be helping to save them! However, as you watch these fascinating animals it is important to keep in mind that these rays are sensitive to disturbance. Unmanaged human interactions with manta and devil rays will lead to negative impacts on the local populations as the number of these interactions increase.

Many dive operators regionally have developed guidelines for encounters with mobulids, especially manta rays. These guidelines help to ensure disturbance to the animals is minimised. They aim to create a sustainable approach to mobulid ray tourism, ensuring these special sites remain healthy for many more years to come. To make sure that the most effective regulations are implemented globally, the Manta Trust has taken a systematic approach to addressing this need by developing a Best Practice Code of Conduct based on long-term scientific studies and information pooled from previous guidelines.

Manta rays are marine cash cows, generating US$73 million annually through direct revenue generated from dive and tourism activities globally. Tourism has opened up many new opportunities to exploit and profit from our oceans; tourism can and should be used to protect them too. At Hanifaru Bay (below) SCUBA diving is now prohibited to protect this important manta aggregation site.

Code of conduct

Over a decade of research from sites all across the globe has provided our team with great insights on manta and devil ray tourism interactions. Based on these studies, a tourism Code of Conduct for interacting with manta rays is summarised below and is available for download from the Manta Trust's *SwimWithMantas* website **www.swimwithmantas.org**, along with educational videos in multiple languages. The Code of Conduct and associated documents on this website refer not only to in-water behaviour by divers and snorkellers, but also include recommendations for vessels approaching and departing manta aggregations, and key points to include in the briefing by crew and operators prior to the manta encounter experience. More broadly, shark and ray tourism operators can also find on the Manta Trust's main website **www.mantatrust.org** a detailed guide: '*Responsible Shark and Ray Tourism - A Guide to Best Practice*', created by the Manta Trust in partnership with WWF and Project AWARE.

Although every site is different, on a broad level there are several steps you can take to minimise the negative impacts on these graceful giants (including the larger devil ray species). Most notably, interactions at manta cleaning stations will differ from those at feeding aggregations, so the Code of Conduct recommends different practices for each. Cleaning stations are often located at prominent reef outcrops, on reef crests, or around coral heads which are home to small specialised cleaner fishes. Manta rays visit these cleaning stations to have parasites cleaned from their bodies. Manta feeding aggregations often occur at predictable locations where ocean currents and the reef geography concentrate the rich planktonic creatures the mantas feed on.

The simple diagrams on the following pages are designed to clearly lay out the key guidelines for interacting with manta rays, both at cleaning stations and during feeding events, either while snorkelling or while SCUBA diving.

Millions of people around the world each year collectively spend billions in search of their own close encounters with marine animals, like these free-divers at a manta ray feeding site in the Maldives (right). Just a few decades ago rays were generally perceived as dangerous sea monsters.

HOW TO SWIM WITH
MOBULID RAYS

By following this Tourism Code of Conduct, you will avoid disturbing the mantas you encounter. At the same time you will increase your chance of having a life-changing experience with these gentle giants.

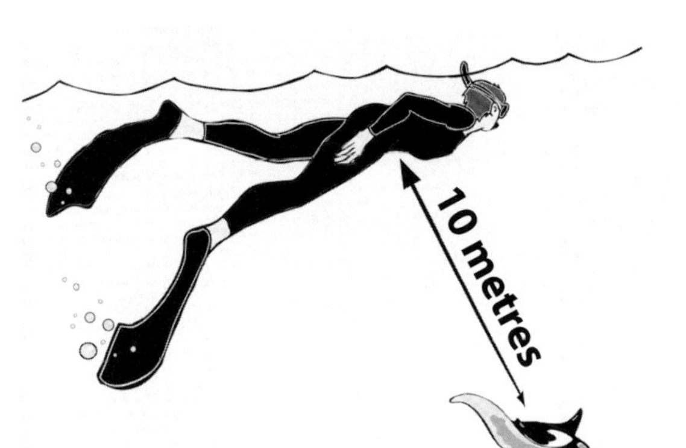

STEP 1

Enter the water quietly and calmly, no closer than 10 metres/33 feet from the manta ray.

STEP 2

Keep your **fins below the water's surface** when swimming. Splashing and noise can scare mantas away, so you want to approach as quietly as possible.

STEP 3

Do NOT approach closer than 3 metres/10 feet. Instead, remain still and let the manta come to you.

STEP 4

You should **approach the manta from their side**, giving them a clear path ahead.

STEP 5

As the manta swims past you, **do NOT chase after it!** You will never catch up with a manta anyway, and will likely scare it away in the process.

STEP 6

Do NOT touch a manta ray. You will ruin the encounter, and may receive a fine depending on local laws.

STEP 7

For scuba divers only.

If you are diving with mantas, you will most likely be encountering them on a cleaning station. These are important sites for manta rays.

During the encounter, **remain at the side of the cleaning station. Do NOT swim onto the main cleaning area.**

STEP 8

For scuba divers only.

Keep low and hover close to the seabed, but **be careful not to damage the reef** beneath you. Depending on the dive site, you may need to stay in an area designated for divers.

STEP 9

For scuba divers only.

When a manta swims towards you, **do NOT block their path as they swim overhead.** Stay low, and stay where you are.

STEP 10

Be sure to **follow any extra rules**, laws and regulations that may be specific to the manta site you're visiting.

To watch a film version of this guide, and learn more about sustainable manta tourism, visit:

www.SwimWithMantas.org

APPENDICES

Data collection protocols

Data collection of mobulid rays across the world is essential if we are to learn more about these threatened species and, in so doing, better conserve them. Therefore, we encourage readers to keep good records of your observations. As citizen scientists or local researchers, you are key contributors to our knowledge of these species. We do, however, note that data collected from living animals versus data collected from dead animals (fishery landing sites, strandings, accidental entanglements, etc.) are two very different scenarios. For in-water photo ID sightings of manta rays, refer to the Citizen Science section of this guide book for further information on how to contribute images and sightings data to our *IDtheManta* global database. Detailed below are protocols for collecting data on live and dead specimens of **any** mobulid species. Any images and data collected can be submitted to the Manta Trust at **info@mantatrust.org** and will be passed on to the scientists researching the species in question in the region where the data was collected. We always endeavour to provide feedback on any data submitted to us as soon as possible.

Data collection from live specimens

Note: *please adhere to the Manta Trust's Code of Conduct while interacting with wild animals;* **www.swimwithmantas.org.**

Where possible, the following should be collected:

1. **Name** and **contact information** of data collector.
2. **Date, time** and **location** of encounter; including habitat type, dive site, water depth, etc.
3. **Images of the individual** (if an underwater camera is available). Ideally images should be taken from above, side-on (especially head), and underneath (including spot patterns if any, and genital region.
4. **An estimate of size**; disc width (DW = wing-tip to wing-tip).
5. **Species identification**; using this field guide, determine the species of the observed individual.

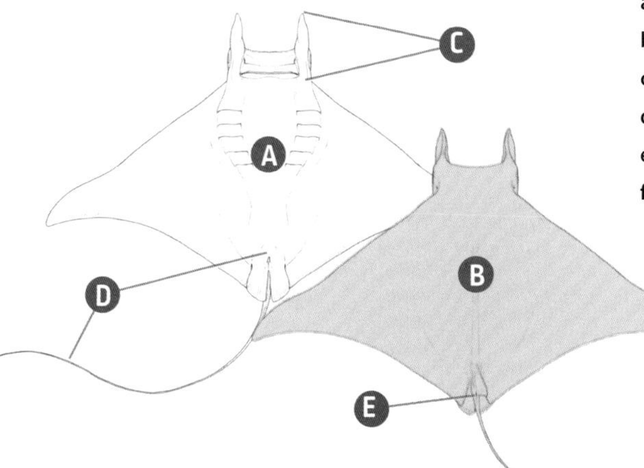

Data collection from dead specimens

Note: *any specimens encountered at fishery landing sites, either as target or bycatch fisheries, or that may have washed up dead on shore, may be recorded. However, any data collection must not incentivise fisher folk, or others, to target or collect additional specimens for scientific purposes. Where applicable, permission must also be obtained from the relevant local authorities before data is collected.*

Where possible, the following should be collected:

1. **Name** and **contact information** of data collector.
2. **Date, time,** and **location** where specimen was recorded.
3. **Specimen number.** This is only required if multiple individuals are encountered (especially if images of multiple specimens are taken).
4. **Photograph** of each encountered individual. Where possible, images showing each of the following characteristics should be recorded for ***each*** specimen:

a) Entire ventral surface including any markings, if present
b) Entire dorsal surface
c) Head and mouth
d) Tail, pelvic fin, and genital region
e) Base of tail showing presence or absence of spine
f) Spiracle, eyes, and cephalic fins.

Label each photograph so it can be matched to the relevant specimen.

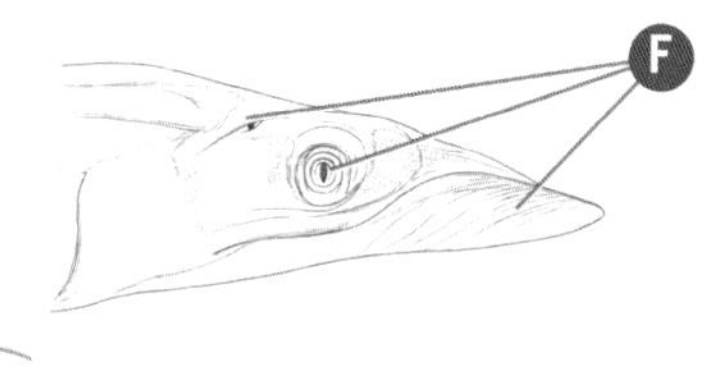

Data collection from dead specimens (contd.)

5. Measurements:

***Note:** it is critical that the measuring tape is **not** bent, and does **not** follow the contours of the body. In other words, the width or length of an animal should not be affected by how 'fat' or 'thin' it is. The tape should be held straight from point to point.*

a) **Disc width (DW):** length in centimetres (cm) from pectoral fin tip to fin tip in a straight line.

b) **Disc length (DL):** length in centimetres (cm) from the central point of the top jaw (*not* the cephalic fins), to the base of the ray's pectoral fins (*not* the pelvic fin or claspers).

6. Sex and maturity:

a) **Sex (male or female)** – check for the presence of claspers for males, and take a photograph (as shown opposite) to help judge maturity. For more information on sexing mobulid rays, see page 64 of this guide.

b) **Maturity (juvenile, sub-adult, adult)** – maturity of females is difficult to determine unless a pregnancy, or mating scars (see pages 46–49 of this guide), are observed. For males, maturity can be judged based on size and extent of calcification of the claspers. In general:

Claspers that extend beyond the pelvic fins and appear, or feel, fully calcified, can be classified as **mature males**.

Claspers that extend up to the pelvic fin, and are partially calcified (flexible), can be classified as **sub-adults**.

Juvenile males have undeveloped claspers that are not calcified.

7. Species identification: using this field guide, determine the species of the observed individual.

8. Additional information:

Catch location – distance from shore and, ideally, GPS coordinates.

Catch method – determine the fishing method used (e.g. gill nets, longline, purse-seine).

Other information – this should include data such as unusual markings (e.g. shark bites or scars) on the specimen, presence of remoras, information such as the boat registration number, the quality of the landed specimen (i.e. was the meat fresh or rotten?), if the ray was killed or alive when brought onto the boat.

Example of data log entry form:

Manta TRUST

Data Entry Log Sheet for Mobulid Rays

Date:

Location:

Sheet Number:

Collected by:

Specimen Number	Picture Number	DNA Sample Number	Species	Gender (M/F)	Maturity (Yes/No)	DW (cm)	BL (cm)	Catch Location and Method	Additional Info (boat registration, shark bites, remoras, scars etc.)

Collecting tissue samples from dead specimens

Note: *the collection of tissue, or other, samples for scientific purposes such as genetic studies should be conducted in accordance with local and national regulations. These vary from country to country. The following instructions are intended for the collection and preservation of samples to be used primarily for genetic purposes (extraction of DNA) and should only be undertaken from dead specimens with the necessary permits.*

Equipment

Ethanol: it is essential that **pure lab-grade Ethanol** (>70%, and ideally 99% for long-term storage) is used to store samples. Over-the-counter ethanol contains ***methanol***, a chemical added to make ethanol poisonous for human consumption. However, methanol degrades tissue samples, making genetic tests impossible. In order to avoid this, it is essential that ethanol is purchased from a recognised laboratory or pharmaceutical company that specialises in ethanol for laboratory/genetic purposes. Ensure that the ethanol is not more than two years old from date of production as it decreases in strength. Make a note on storage tubes when ethanol was dispensed into them.

Storage tubes: 2ml screw-cap tubes (with an O-ring to prevent leakage) to store the samples in. These can be purchased at laboratory supply stores (usually similar to the ethanol supplier) or through universities.

Scalpels: scalpel blades (surgical blades) with a scalpel handle are best suited to obtain a sample. They can be purchased at most pharmacies.

Sampling kits

If you have the necessary permits along with access to dead fishery specimens but are unable to obtain the necessary vials or ethanol, please contact the Manta Trust **info@mantatrust.org** for a complete sample kit. *Do not* use ethanol you are unsure of, as you may damage the samples, making them worthless for genetic studies.

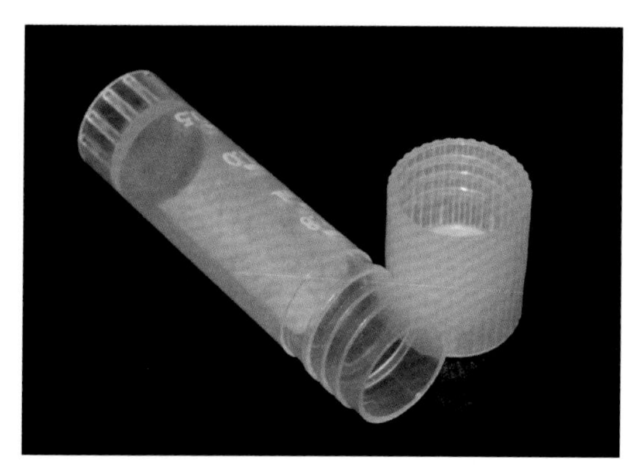

2ml screw cap storage tubes, with an O-ring to prevent leakage, used to store the tissue samples.

Procedure

Collect all possible biological data from the specimen as stated in the above sections. Each tissue sample should correlate with the respective data entry and images of the specimen.

Cut a thin strip of meat/muscle tissue from the specimen. A small piece of tissue is best, but a piece of the tail will also suffice. This should be around 1–2cm long and should fit inside the 2ml storage tube (pictured opposite).

Completely immerse the sample in ethanol inside the tube. The sample should ***not*** be too large and the ratio of ethanol should be equal to or greater than the size of the tissue sample. **This is really important:** samples that are too big will not preserve properly in the amount of ethanol contained within a 2ml tube.

Clearly **label** and **number** the tube using a ***permanent*** marker pen:

Name of specimen – e.g. *M. mobular*
Date of collection – e.g. 10/05/2017 (DD/MM/YYYY)
Sample number – e.g. 523

Note: *even a permanent marker can be erased by ethanol. Ensure that ethanol does not leak out of the tubes (use caps with O-rings) and store the tubes in a chronological order, along with a copy of collected sample data on a backed-up spreadsheet.* ***Duplicate the sample number*** *by also marking the storage tube lid as well as the side of the tube.*

Place the sample tube upright in a tube storage rack or cryogenic box.

Store samples in a cool, dry location (away from direct sunlight). Ideally keep samples in a freezer (below -4°C or if possible, ideally at -80°C). However, if you live in a region with frequent power-failures resulting in the repeated freezing and thawing of samples, store samples in the refrigerator instead or outside in a cool, dry location.

If samples are not being shipped immediately, change the ethanol (with pure-lab grade ethanol) in the tubes once after 3–7 days of collection. This ensures that all the water extracted from the sample is removed and replaced with fresh ethanol. For information on other tissue sampling procedures (stable isotope, heavy metals, etc.), or if you have any further questions about our data collection protocols, please don't hesitate to contact the Manta Trust at **info@mantatrust.org**.

Law enforcement and trade monitoring guide to the gill plates of mobulids

All mobulids (manta and devil rays; Mobulidae) are listed on Appendix II of the Convention on International Trade in Endangered Species of Wild Fauna and Flora (CITES). Found throughout the world's tropical and temperate oceans, mobulid rays are large animals with few natural predators. Their biological characteristics, including their late maturity, long gestation periods, and low fecundity (giving birth to one single pup), make them highly vulnerable to directed or bycatch fisheries.

Gill plates (prebranchial appendages)

All mobulids are filter feeders, using their huge mouths and modified gill plates to strain plankton and small fishes from the water. Each mobulid ray has five pairs of gill arches, each of which is encircled internally by a ring of feathery gill lobes known as prebranchial appendages or gill plates.

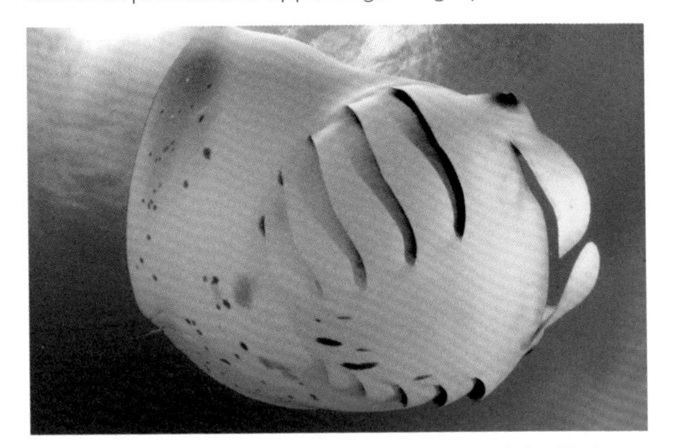

Five pairs of gill slit openings of a Reef Manta Ray *Mobula alfredi*.

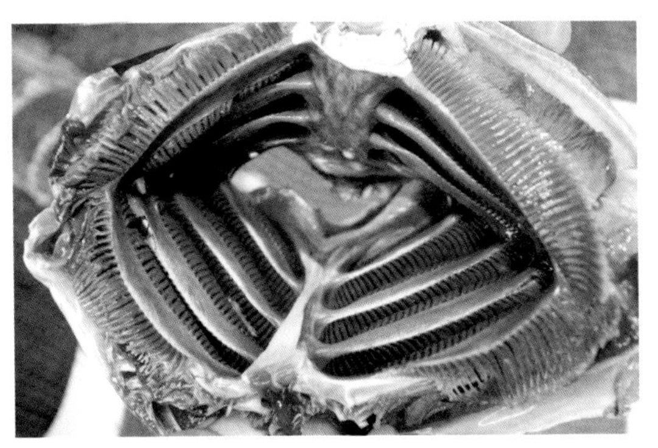

Feathery prebranchial appendages encircle the gill arches inside the mouth of the mobulid ray.

Gill plate trade

Although gill plates from five different species of manta and devil rays have been found in the gill plate trade, only the mantas and two largest mobula species (*Mobula tarapacana* and *M. mobular*) are regularly traded. Gill plates from the two species of manta rays can be visually differentiated from any devil ray species, and the gill plates of these two most-traded devil ray species can also be easily distinguished from one another.

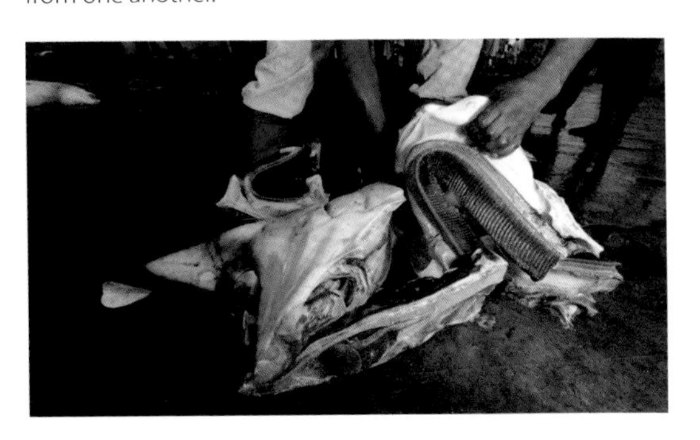

The gill plates of manta and devil rays are used in Asian medicine. Growing demand is driving a global fishery for these species. When the gill plates are removed from the dead animals (left), they are cut in half before being dried (above) and then shipped to the point of sale.

Effective enforcement and monitoring of international trade in these CITES-listed species will be enhanced through the ability to easily identify these gill plates, and the ability to distinguish between the gill plates of the different manta and devil rays being traded. This guide is therefore intended to help fisheries, enforcement, and customs personnel in the identification of mobulid gill plates. Definitive DNA tests are also available to confirm visual identification if needed for prosecution or verification purposes.

Above: customs officials in Indonesia with hundreds of seized manta ray gill plates which were being traded illegally.

Left: gill plates from the Sicklefin Devil Ray *Mobula tarapacana* are known as 'flower gills' in the gill plate trade.

Gill plate features

There are three key features that can be used to easily identify each gill plate type:

1. Gill plate size
measured as the total length of the traded gill plate.

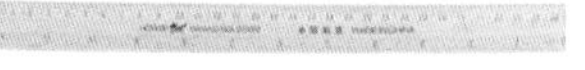

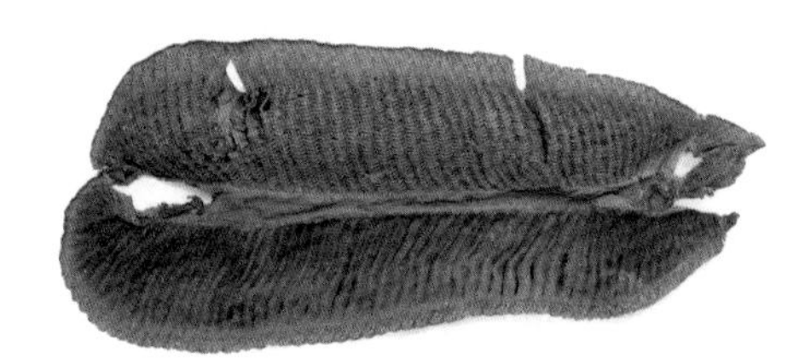

2. Gill plate colour
uniform (above) or bicoloured (below).

3. Gill plate lobe edging
smooth (above) or separated/bristled (below).

Key to visual identification of the most commonly traded mobulid ray gill plates

Question 1:

Is the gill plate longer than 30cm and uniform brown/black in coloration?

YES = **manta rays**

NO 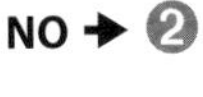

Manta rays *Mobula birostris* and *M. alfredi*

1. **Size** = medium/large (usually more than 30cm).
2. **Colour** = uniform brown/black (rarely white).
3. **Lobe edging** = smooth.

Question 2:

Does the gill plate have a white central coloration and smooth lobe edging?

YES = **Sicklefin Devil Ray**

NO = **Spinetail Devil Ray**

Sicklefin Devil Ray *Mobula tarapacana*

1. **Size** = medium.
2. **Colour** = bicoloured (white central).
3. **Lobe edging** = smooth.

Spinetail Devil Ray *Mobula mobular*

1. **Size** = small/medium.
2. **Colour** = bicoloured (white edging).
3. **Lobe edging** = jagged.

Conclusions

Mobulid gill plates can be easily identified using this simple visual identification guide.

The size, colour patterning, and lobe edging of the gill plates can be used as an effective and easy indicator to determine the species of origin.

Spinetail Devil Ray, *Mobula mobular*, displaying courtship behaviour, Ari Atoll, Maldives.

Glossary

Allele Each of two or more alternative forms of a gene that arise by mutation and are found at the same place on a chromosome.

Allopatric Describing animals or plants, especially related species or populations, that occur in separate, non-overlapping geographical areas. Opposite of **sympatric**.

Bathypelagic zone Also known as the midnight zone, this is the part of the open ocean that extends from 1,000 metres (3,300 feet) below the ocean surface to a depth of 4,000 metres (13,000 feet).

Cephalic fins Literally 'head fins', these horn-like projections on either side of the head of manta and mobulid rays are used to funnel prey into the ray's mouth.

Claspers The pair of appendages under the abdomen of a male shark or ray used to transport semen into the female's cloacal opening during copulation.

Cloaca The cavity at the end of the digestive tract for the release of both excretory and genital products in vertebrates (except most mammals) and certain invertebrates.

Convergent evolution In evolutionary biology, the process whereby organisms that are not closely related independently evolve similar traits as a result of having to adapt to similar environments or ecological niches.

Coral bommie A stand-alone coral structure that can be as small as a beach ball or as big as a car. It is a structural stepping stone, providing a refuge or home for many marine organisms in the maze of reef structures.

Ecological niche The role and position of a species within its environment: how it meets its needs for food and shelter, how it survives and how it reproduces. A species' niche includes all its interactions with the biotic and abiotic factors of its environment.

Indicator species An organism whose presence, absence or abundance reflects a specific environmental condition. Indicator species can signal a change in the biological condition of a particular ecosystem and thus may be used as a proxy to diagnose the health of an ecosystem.

Lateralisation The localisation of function on either the right or left side of the brain.

Lek A traditional place where males assemble during the mating season and engage in competitive displays that attract females.

Leucism The condition in which an unusually low concentration of melanin and other pigments occurs in the skin, plumage or pelage of an animal.

Melanism The condition in which an unusually high concentration of melanin and other pigments occurs in the skin, plumage or pelage of an animal.

Mutualism The relationship between two species of organisms in which both benefit from the association.

Oviduct The tube through which an ovum or egg passes from an ovary.

Pelagic Relating to or living in the open sea.

Philopatry The tendency of an organism to stay in or habitually return to a particular area. The causes of philopatry are numerous, but natal philopatry, which is when an animal returns to its birthplace to breed, may be the most common form.

Phylogenetics In biology, the study of the evolutionary history and relationships among individuals or groups of organisms.

Plankton The diverse group of organisms (plants and animals) that live in the water column of large bodies of water and cannot swim against a current.

Rete mirabile Vascular bundle comprising exceptionally long, thin blood vessels with coiled or looped form.

Seamount A peaked, underwater mountain that rises at least 1,000m above the ocean floor.

Sexual dimorphism The distinct difference in size or appearance between the sexes of an animal in addition to the sexual organs themselves.

Site fidelity One type of philopatry is breeding philopatry, or breeding-site fidelity, which is when an individual, pair or colony returns to the same location to breed, year after year.

Speciation The formation of new and distinct species in the course of evolution.

Spiracle An external respiratory opening, as in the pair of vestigial gill slits behind the eye of a cartilaginous fish.

Symbiosis The interaction between two different organisms living in close physical association, typically to the advantage of both.

Sympatric Describing animal or plant species or populations that occur within the same or overlapping geographical areas. Opposite of **allopatric**.

Taxonomy The branch of science concerned with classification; the process or system of describing the way in which different living things are related by putting them in groups.

Thermoregulation The maintenance of a constant internal body temperature independent of the environmental temperature. The state of having an even internal temperature is called homeostasis.

Vestigial Describing the very small remnant of something that was once greater or more noticeable.

Index

Page numbers in **bold** refer to species accounts